ROUTLEDGE LIBRARY EDITIONS:
SOCIAL AND CULTURAL GEOGRAPHY

Volume 9

T0264817

REMAKING HUMAN GEOGRAPHY

REMAKING HUMAN GEOGRAPHY

Edited by
AUDREY KOBAYASHI
AND SUZANNE MACKENZIE

Routledge
Taylor & Francis Group

LONDON AND NEW YORK

First published 1989 by Routledge

2 Park Square, Milton Park, Abingdon, Oxon OX14 4RN
711 Third Avenue, New York, NY 10017, USA

Routledge is an imprint of the Taylor & Francis Group, an informa business

First issued in paperback 2016

British Library Cataloguing in Publication Data
A catalogue record for this book is available from the British Library

ISBN 978-1-138-98506-3 (pbk)
ISBN 978-0-415-73332-8 (hbk)
ISBN 978-1-315-84849-5 (ebk)

Publisher's Note
The publisher has gone to great lengths to ensure the quality of this reprint but
points out that some imperfections in the original copies may be apparent.

Disclaimer
The publisher has made every effort to trace copyright holders and would
welcome correspondence from those they have been unable to trace.

REMAKING HUMAN GEOGRAPHY

Edited by

Audrey Kobayashi

Suzanne Mackenzie

Boston
UNWIN HYMAN
London   Sydney   Wellington

Unwin Hyman, Inc.
8 Winchester Place, Winchester, Mass. 01890, USA

Published by the Academic Division of
Unwin Hyman Ltd
15/17 Broadwick Street, London W1V 1FP, UK

Allen & Unwin (Australia) Ltd,
8 Napier Street, North Sydney, NSW 2060, Australia

Allen & Unwin (New Zealand) Ltd in association with the
Port Nicholson Press Ltd,
Compusales Building, 75 Ghuznee Street, Wellington 1, New Zealand

First published in 1989

Library of Congress Cataloging-in-Publication Data

Remaking human geography / edited by Audrey Kobayashi &
Suzanne Mackenzie
p.   cm.
Bibliography: p.
Includes index.
ISBN 0-04-445325-6
1. Anthropo-geography—Philosophy.   2. Anthropo-geography
—Methodology.   I. Kobayashi, Audrey Lynne.   1951–
II. Mackenzie, Suzanne.
GF21.R46 1989
304.2—dc19

British Library Cataloguing in Publication Data

Remaking human geography.
1. Human geography
I. Kobayashi, Audrey II. Mackenzie,
Suzanne   304.2
ISBN 0-04-445325-6

Typeset in 10 on 11 point Bembo by Columns of Reading
and printed in Great Britain by Billing & Sons Ltd, Worcester

Foreword

This book is concerned with developing a dialogue between humanism and historical materialism in human geography. It seems to me that its contributions differ from those made by participants in the debates which took place in the 1970s in a number of important ways. These emerge with great clarity from the essays which follow, but it might be helpful to signpost some of them in advance (though there are obvious dangers in doing so: there are differences between the contributors which ought not to be glossed over in any genuine debate).

First, a programme of this kind does not involve a direct mapping of philosophy or social theory into human geography. In the past, I suspect, there was a tacit assumption that most of our problems had to be resolved on some privileged terrain outside the discipline and then projected on to it. On this reading, geography was at best parasitic on developments taking place elsewhere. The situation has now changed, and there is an increasing recognition of the importance of concepts of place, space and landscape across the whole field of the humanities and the social sciences. Their incorporation does not leave the traditional orthodoxies unchanged – far from it – but within human geography (if we can still speak within the usual disciplinary enclosures) humanism can no longer be portrayed as a parochialism in which the day-to-day lives of particular people in particular places is the single focus of study, any more than historical materialism can continue to be criticized as a brute globalism in which the specificities of history and geography are read off from the universal logic of the (capitalist) mode of production. Both traditions have moved beyond those simple reductions towards a recognition of the complexity of human geographies: towards the realization that, as Mann puts it, 'societies are much *messier* than our theories of them'.

Second, it follows that neither humanism nor historical materialism are exercises in pure abstraction. Some of the early writings (on both sides) probably deserve that description: too many appeals to 'the' human subject or to 'the' logic of capital. But no more. Humanistic geography is increasingly concerned with

making sense of the dynamics of concrete 'language-games' and
not, as Geertz once so pungently put it, with 'discovering the
Continent of Meaning and mapping its bodiless landscape'.
Historico-geographical materialism, too, is much more attentive
to the variability of capitalism across time and space and the
emergence of distinctively new regimes of accumulation. To say
this is not to imply that either of them is inimical to theoretical
reflection or to the construction of theoretical systems – all of
these essays demonstrate the connective imperative between the
theoretical and the empirical – but it might well require us to re-
think what we mean by (and expect of) theory itself: to re-cast the
relations between what some commentators see as Grand Theory
(or not-so-grand theory) and 'local knowledge'.

Opening up that wider conversation has required a third
change: a movement beyond the old absolutisms which all but
paralysed previous discussions. This is more than an argument
about recognizing the plurality of traditions within both humanism
and historical materialism (important though that is); it is also
about recognizing the cross-currents between them. Let me give
two obvious examples. Contemporary versions of humanism
need to be distinguished – and certainly more clearly distinguished
than is usually the case – from liberal individualism and its
characteristic emphasis on the autonomy of the individual. It
would not be difficult to identify a series of developments, all of
them located within a broadly conceived humanism, which
accentuate the 'boundedness' of human agency and seek to
disclose the various systems and structures which enter into the
constitution of social life. Equally, historical materialism cannot
be confined to its largely structural versions; these were never as
constricting as most of their critics seemed to think, but in any
event there are now versions of 'rational choice' or 'game-
theoretic' Marxism which claim that all social phenomena are
explicable only in terms of individuals and attempt to expose the
'microfoundations' of historical materialism. It should go without
saying that developments of this kind need careful scrutiny, and I
would not want to endorse either of these particular examples; but
taken together they do suggest that critical examination (and
reconstruction) ought not to mean a narrowly sectarian exegesis.
Of course there are differences between intellectual traditions, and
I think it unlikely that these can be resolved by appeal to any
supra-disciplinary court; but I also think it intrinsically implausible
to assume that any one perspective has a monopoly of understanding
or insight. I realise that this raises the spectre of relativism – even
nihilism – but in so many ways the essays in this collection point

to a movement beyond both relativism and absolutism. It is not a matter of accepting that 'anything goes', that one interpretation is as good as any other – it doesn't and it isn't – but rather of recognizing the creative tension which emerges through the mediation of different frames of reference.

That said, choices have to be made, of course, and this has the liveliest of practical implications. I see this as the fourth moment; both humanistic geography and historico-geographical material-ism have become more participatory projects. I don't think I was alone in detecting a rasping elitism in some of the early formulations of humanism, in its more conservative forms at least, and within Marxism too there was, at times anyway, an unmistakable presumption that the task of radical intellectuals was to plot the collective destiny of other people. But the essays in this book demonstrate a greater sensitivity towards the deeper meaning of 'making history': towards the interpenetration of human agency and social structure. And in doing so they also show exactly what is at stake in remaking human geography.

DEREK GREGORY

For our parents, who gave us childhood
in the interior of British Columbia

Acknowledgments

This collection was inspired by a set of sessions organized around the themes of humanism and historical materialism at the Annual Meeting of the Canadian Association of Geographers in Nanaimo, British Columbia, in 1984. We wish to thank all the people who participated in and attended those sessions. In keeping with the spirit of contextual approach that provides the main theme of the papers in this book, the editors wish to acknowledge the important roles played by their respective educational experiences with their undergraduate years at the University of British Columbia and Simon Fraser University respectively. We would also like to thank those who provided encouragement and constructive criticism in the early stages, among others, Michael Dear, Ruth Fincher, Peter Jackson and Mark Cohen, who helped us through the editing process. For their support, we thank Alan Nash and Mark Rosenberg, friends, geographers and husbands. The Departments of Geography at Carleton and McGill have provided much administrative assistance. We especially thank Marie Raymond at Carleton and Angela Mansi at McGill. The volume has been a collective effort by all contributors. We thank them for their patience and cooperation. Special acknowledgement is due to John Bradbury, whose paper here is being published posthumously. Until a few days before he died, John played a very significant role as friend, colleague and mentor, and we owe much of both the inspiration and the hard slogging to his encouragement and suggestions. Finally, the editors take full responsibility for errors and omissions.

AUDREY KOBAYASHI & SUZANNE MACKENZIE

Contents

Section III: Directions 185

People make their own history, but they do not create it at will or under conditions of their own choosing; rather, by circumstances directly encountered, given and bequeathed.

Die Menschen machen ihre eigene Geschichte, aber die machen sie nicht aus freien Stücken, nicht unter selbstgewählten, sondern unter unmittelbar vorgefundenen, gegebenen und überlieferten Umständen.

Marx, Karl 1946. *Der achtzehnte Brumaire des Louis Bonaparte.* Berlin: Verlag JHW Dietz Nachf.

Introduction: humanism and historical materialism in contemporary geography

AUDREY KOBAYASHI
& SUZANNE MACKENZIE*

This is a book about some of the ways we do geography, and especially about how geographic research can provide understanding of some of the issues that dominate the current agenda of citizens, academics and policy makers. We are concerned to understand how social processes have motivated new theoretical and methodological debates in geography, how we have responded to these, and what have been some of the outcomes of our response.

We live in the world, among our data. This truism sets the basic condition for the fact that we also analyse the world, in a disciplined and disciplinary manner. We group, exclude and emphasize using criteria that are systematically learned, but of which we are seldom wholly conscious. In so doing, we change the world, alter our database, not only in restructuring the views of our students, colleagues and the amorphous public, but also in effecting environmental outcomes of our own actions and ideas. We are concerned, therefore, to discover not only the kind of world in which we are living, but also to discover how we as geographers inhabit, reproduce and change that world, and how we understand and control the process of change. Our work is active. Whether we consciously incorporate 'dynamism' into our frameworks, our work is active and dynamic, and necessarily, constantly changing. New issues, emerging forms of social organization and the decline or alteration of existing ones coerce

*The two editors are equally responsible for this book. The names are listed in alphabetical order.

us into new ways of understanding, and we adjust our theoretical perspectives to meet the exigencies of social change.

Sometimes, if the alterations are significant enough and the institutional climate sympathetic, we adopt new perspectives. This has been the case in geography, with the exploration and incorporation of two perspectives that can broadly be referred to as humanism and historical materialism. Both emerged from the social unrest and the academic reorientation that occurred during the late 1960s and early 1970s. Both have now passed through the zealous and often irascible pioneer stages, to a period of more mature and critical discussion of methodology and the production of substantive empirical work. Now, we are reaping the fruits, bitter and sweet, of the social ferment that coalesced two decades ago, and the harvest reflects both the processes we must analyse and the analytical tools at our disposal. Better equipped than we were in the earlier period, we are also confronting difficulties that are in part a legacy of the ways we acquired this analytical equipment.

This volume was inspired by the recognition that the contributions of historical materialism and humanism, which we see as complementing each other in many ways, have often opposed each other. The time has come for urging both out of their parallel courses, in an examination of the possibility for both dialogue and reconstruction. This should be undertaken, however, not merely out of a spirit of *rapprochement*, but because the philosophical, methodological and empirical work in humanism and historical materialism indicates areas where mutually beneficial discussion can take place. In attempting to take these developments into account this book, like all books, is both a reflection of the historical period in which it was produced and an attempt to direct the course of future developments. By way of introduction to the chapters, the following discussion attempts to illustrate some of the ways in which our academic concerns can be explicitly situated within a social context.

THE SOCIAL CONTEXT

As the recent period of global 'recession' gives way not to a stronger and purified capitalist equilibrium, but to societies undergoing fundamental economic and social 'restructuring', social scientists are facing, for perhaps the second time since the late 1960s, a set of social changes which appear to demand analytic changes.[1] In the 1960s and early 1970s, we were taken aback by that wide range of protests and problems which shattered the weakening idea of social equilibrium and which coalesced in a variety of movements – tenants and community movements, ethnic and racial movements and the women's liberation movement – now seen as having had an important effect in bringing about permanent social change. They appeared to be expressions of new, or at least hitherto unarticulated, problems in society, and to be making unprecedented and interconnected demands upon the city, the region and, by implication, upon our analyses of the city and region. There followed a period of critical self-examination within the environmental disciplines – geography, planning, architecture, urban and regional studies – during which our data escaped the lumbering computers of the period, broke through the boundaries of our conceptual categories and took to the streets, the council chamber and even the classroom.

In retrospect, we can see that these changes were part of an ongoing process which is characterized by, among other things, extensive economic and social 'restructuring' and global forms of communication and cultural practice. But these changes also present us with a paradox. The circumference of human life is both constricting and expanding. On the one hand, a continuous restructuring of economic and social conditions seems to restrict options: efficiency in production is accompanied by declining numbers of jobs; global marketing appears to involve an international homogenization of cultural forms and economic structures. We seem to have few choices and less space in which to create them. On the other hand, for the first time in history we have both the technological and the intellectual means to act as global citizens, and global changes have elicited a variety of means by which we can react to and further open up disparate areas of life to intellectual analysis as well as conscious political action. There appear to be more places, more parts of our lives, in which we can deliberately act and exercise control: our relations within the household and neighbourhood; our patterns of sexuality; our ability to communicate at a local and world scale.

This means that many of us are not only motivated to examine, for example, the alteration of wage employment patterns, shelter groups or paths of movement in cities and regions, but are living out these changed patterns. In our dual career families wearily commuting up and down disintegrating freeways, in our terminal contracts, with our neighbours in gentrified districts, with our political allies in a variety of *ad hoc* and official alliances, we are both inspired and confounded by the questions that social change creates. Perhaps nowhere has this been more evident than in the relations between people and their environments.

Geographers have responded to these conditions by seeking a view of the social totality which contains both people and the environment, and which makes explicit the connections among the apparently disparate ways in which the social world is structured. In this respect, the past decade and a half of geographical enquiry has been part of the continuity established by the positivist tradition when it broke with the particularist and ideographic tradition of regional studies which had dominated geography in the 1930s and 1940s.[2] But we have also broken with the geographic outcome of positivism – spatial science – with its emphasis on abstract relations, and its claim to being value free. We have chosen instead perspectives that make salient the relations of people and environment in context, without casting aside the complexities of this relationship when they could not conform to our models.

In short, we wanted everything. Two decades ago, in the heady days of the 'cultural revolution', when this reassessment began, everything seemed a reasonable thing to ask, both of our own actions and of our conceptual understanding. Some sought to understand the constitution of meanings and values towards the environment, while others recognized the importance of structural links which created and constrained choice in forms of environmental appropriation. We all sought a view of the social totality which included the internal interaction between people and their environment but which did not reduce this to idiosyncrasy, which saw interaction in universal terms, as a base from which to develop analytical generalizations.

But we required more than a new concept of the social totality. We also required a new concept of social change. Although change had been embedded in the earlier regional tradition, it was an evolutionary concept of change, change which appeared to move in slow, albeit graceful, sweeps across a people and a (usually rural) landscape, change which left distinct traces. We appeared to be living in a *revolutionary* period, where change was rapid. Buildings fell, countries divided, spaces for previously inarticulate and formless groups (women, gays, blacks, international refugees, and victims of family violence and psychiatric treatment) were created overnight. The rhythm which had governed the regionalists' evolutionary change had given place to an often discordant but exhilarating beat.

Change had also been present in spatial science, but it was a flat and personless change. One could assume there were real people creating those patterns, flows and hierarchies, or at least making the phone calls or shipping the goods which generated movement, but it was difficult to put faces on them. They acted, but apparently without feeling, motivation, or the consideration of alternatives. We were, and we saw around us, quite different kinds of actors, people who not only had faces – often black or female ones – but people involved in conscious, often angry attempts to change the constraints under which they acted, to change the flows and patterns which created these constraints. Above all these people were attempting to act in ways which could not even be seen, let alone understood, within the concept of change that rested on the twin pillars of a theoretical equilibrium and methodological duplication of results. Most of us not only did not expect to find current trends extended and results of previous experiments duplicated, we did not want to. Nor did most of the people we worked with, interviewed, or peered at through the foggy lens of officially collected statistics. We were social agents, however clumsy and inarticulate, and we needed to understand, both as academics, and more urgently, as women and men living in society, what it was that we were creating. We suspected, too, that in the process of creating a new society we were somehow creating new forms of social groups, new alliances which dissolved old categories.[3] Perhaps we were even creating new kinds of people.[4]

It was with this set of requirements – concepts of the social whole, of the synthetic relation between people and the environment, and of social change – that we began to explore alternative ways of looking at the world which incorporated the rapid and unratified creativity of newly coalescing groups. This exploration began with what has been called in retrospect the 'relevance debate'. This debate, which is given specific recognition (at least for Anglo-Canadian-American geographers) in the journal *Area* in the early 1970s and in the arrival upon the scene of the irreverent *Antipode* in 1969, brought to a close a period of unexamined theoretical and institutional optimism. Geographers attempted to make themselves and their discipline more relevant either to policy makers, as a means of ensuring the survival of the discipline, or to disadvantaged groups in society, as a way of making geography more socially relevant and ensuring it had a place in the 'new society' which was emerging out of the shifting protests and alliances being formed around us.[5]

Some geographers, most of them of an age to have relatively little stake in geography as an intact discipline and a way of making a living and a great stake in changing the social reality within which their vision of geography could be placed, extended the relevance debate to become advocates. Geographers must, they argued, reorient the discipline to solve problems defined not by society's power groups, nor even by socially conscious geographers, but by the disadvantaged groups in society itself. They undertook geographic expeditions, living in the communities they studied and acting as advocates for other residents.[6] Advocacy geographers changed the definition of what constituted research, claiming that 'research is not complete until implementation is achieved' (Stephenson 1974).

However effective advocacy may have been in challenging the nature and social role of the discipline, it was criticized, often by members of its expeditions, as leading to 'piecemeal' analysis and solutions, and as confining geographers' action and analysis to the study and solving of small scale problems, thus providing only adjustments to the given system. More practitioners argued for a new theoretical base for geography, a 'paradigm shift' which would allow the combined understanding of underlying social forces and strategic action.[7] In response, the underpinnings of our discipline came to be questioned more and more explicitly through two major alternative paradigms: humanism and historical materialism.[8] The former was satisfying because it centred on the importance of human agency in human environmental relations, it allowed us to 'reorient science and knowledge in human terms' (Relph 1970, p.193). The latter was satisfying because it was, in Peter Saunders' terms, 'a science of connections', one which argued that society was a structural unity, and which was explicitly concerned to understand and to change the structural relations which defined the nature of social production and reproduction (Saunders 1981).

Both responses had a concept of the social totality which promised synthesis, and a concept of social change. But these appeared, initially, to be radically different concepts. Humanism seemed to centre people – real, living, complex women and men – who were, potentially, infinitely creative, building out from themselves structures of meaning, value and significance, worlds, landscapes and places that were, in Tuan's words, 'centre[s] of meaning constructed by experience' (1975, p.152). Historical materialism, on the other hand, gave us a connected totality, one which 'exposed' the layers of relations, the complex of constraints and conditions in which the individual was embedded and through which the individual was created as a social being and a member of specific groups within which she or he acted. Not surprisingly, early criticism of humanism by historical materialists focused on its lack of structural rigour, its asocial and transcendental response to the question, 'what is the social whole?'; whereas early criticism of historical materialism by humanists focused on the apparent determinism of agency by structure, the mechanistic and reductionistic definition of the individual by external forces. (See Harris, this volume, for discussion).

But even such criticism was relatively rare in the exploratory period. Geographers were busy mastering the classics of the respective approaches, and applying their ideas to a critique of the still dominant, but decreasingly so, positivism. The hard work of learning to see the world in a new way, the missionary zeal which accompanied each new discovery, the adoption of mutually incomprehensible language, and the apparent urgency of finding the fit between a framework that worked and a set of disciplinary contents which had been produced in quite different contexts, all ensured that initial explorations proceeded upon parallel tracks, and that forays across the tracks were both ill-informed and antagonistic.

This myopia assured that our growing expertise at criticism of positivism was also put to good use in our first attempts at communication. We criticized one another, sometimes to good effect, often with outmoded or dimly conceived ideas of what the 'other side' was doing.[9] There were more and more articles which outlined the basic premises of humanism as it selectively emerged into geography, and historical materialism as it selectively emerged into geography. There were explorations of how we could and would adapt these perspectives to answer questions about people and the environment.[10] All of this rested upon, ignored or misconstrued a pitifully small body of empirical work where people actually tried to *do* humanism and historical materialism.

That impasse has now been overcome in a number of ways, as we continue to face, or face again, a social reality which is raising new questions, and stimulating methodological response. Although the 1960s and 1970s may have been bewildering periods to live through, in retrospect they are becoming more coherent, as are our own earlier responses. For this was a period when social equilibrium was the conventional wisdom, and where, as a consequence, the conventional 'solution' was social engineering. The answers had been found. Science was carefully naturalizing, and where necessary rectifying, minor problems in the adjustment of different groups, classes, genders, races.[11] It was a period where the shattering of the conventional wisdom was thrust upon us by open and evident protest which articulated, if not elegantly, very forcibly and occasionally violently, the existence, even the dominance, of disequilibrium. It was a type of education in which confrontation was accepted as the appropriate, perhaps the only, form of teaching.

A decade later, we have learned to analyse the problems articulated in the earlier period as social injustice, as inequities in the production and distribution of resources, as the need to incorporate meaning and value. The problems have deepened. Yet we look for responsive protest and find what appear to be adaption and acquiescence. Conditioned to a forcibly articulated change, we see instead an unremarked and unresisted disintegration of the conditions of economic and social life, and of political activity. The responses are there, in fact, but they are different from the responses we expected, so different that we often do not see them.

We had looked to humanism or historical materialism to help us understand the protests, and the reality they articulated. We have been forced to restructure these paradigms in order to grasp the quieter responses to what is perhaps a more fundamental alteration of life. The issues discussed in the first section of this book may perhaps require a very fundamental alteration of our theories, of the concepts through which we see life. Furthermore, over the past decade, we have accumulated a decade's theoretical rigour, and more importantly, altered the two paradigms internally as we adopted and adapted them into geography. We have also seen the slow emergence of the outcomes of this philosophical labour in empirical and historical studies we are now carrying out. These have moved us from a position of mutual, often fine-tuned misunderstanding toward a *rapprochement*, even, in some respects, a convergence, if not of theory then at least of mutual respect. This book is an outcome of this *rapprochement*.

The remainder of this introduction is concerned first with the nature of this convergence and second with the continued diversity, as exemplified in the chapters themselves, in the way we do geography.

CONVERGENCE

The way we define issues is in large part a function of what we are able to see, and what we are able to see is in large part a function of our perspectives. None the less, the current form of social and economic restructuring has forced both humanists and historical materialists to look and listen in new ways, and has motivated some form of internal change in our perspectives. Historical materialists have moved toward a concept of history centred on 'human agency' (as Mackenzie & Harris argue explicitly) and humanists have developed a concept of human situated more firmly in what Ted Relph calls 'the geographical imagination' and what other authors (Harris, Barnes, Ley) call, in different ways, the 'context'.

Within Marxism as a whole, there has been a move from the bad old days of economic determinism, in which the social totality was conceptualized as a determining base and a liquid, often flatulent and always determined, superstructure. We have largely moved through the Althusserian moment, which dissolved the social totality into a series of mutable 'instances'. Structuralism focused the social agenda on to the political, but at the same time, by reducing human agents to structurally determined 'bearers of relations', precluded the agent from any power to effect change. We are now moving into a relational concept of the totality, powered by intentional human agents. This internal evolution was powered by people's continued insistence on altering the conditions of their own lives, and their social world, and thereby altering the nature of human.[12]

Humanist perspectives have, over the past decade, undergone an opposite, but complementary transition. Early work in the humanist tradition by geographers such as Lowenthal (1961), Tuan (1971, 1974) and Buttimer (1974), challenged the often complacent assumption that fact could be separated from value and excited a generation of young geographers with its challenge to orthodox epistemology. In retrospect, however, the earlier attempts to use humanist perspectives, especially phenomenology, to establish the link between the material world and lived experience, succumbed to the epistemic fallacy of holding that epistemic traits provided the key to knowledge of the world, and idealism was thus an appropriate substitute for materialism.[13] Knowledge became a reflection of the world rather than an engagement of reality. Not only were such positions philosophically untenable (Samuels 1978), but they precluded the development of a method suitable to either the traditional concerns of social science (Entrikin 1976, Smith 1981) or the emerging concerns to resolve the issue of structure and agency (Cosgrove 1983, Gregory 1978, 1981). More recently, the central lessons of humanism – the anthropocentricity of social life, the need·for reflexivity, and the inseparability of fact and value – have been carried forward in a growing trend toward empirical work that is both philosophically informed and firmly rooted in material existence (Ley 1981, 1983, 1985).

A number of authors note that once we begin to do our work, the apparently irreconcilable differences become tractable, even amenable to synthesis. This spirit allows tolerance as well as (stubborn) eclecticism, and encourages what Ley (quoting Johnston 1987) refers to as 'mutual accommodation'. Cosgrove argues that the tension between humanism and historical materialism 'can be a creative one. . .in teaching our geography as a celebration of the diversity and richness of the natural world and of the places human societies create in their collective attempts to make over that world into the homes of humankind'. Kobayashi insists that the study of landscape, which she assigns a pivotal role in the geographic project, necessitates incorporating the strengths of both perspectives. What unites this work, and perhaps gives cause for optimism, is that the chapters are concerned with theory as a guide not only to understanding *per se*, but to understanding as a guide to action.

DIVERSITY

Talking about convergence throws another difficult issue into the milieu created by the debate on the meaning of humanism (Ley, Harris, Relph) and the orientation of the variety of terms utilized to identify historical materialism/Marxism/political economy. Perhaps it is easier to say what convergence is *not*, for the editors, and for the contributors to this collection. Convergence is not the collapsing of philosophical distinction, but rather the refining of philosophical categories and a drawing away from formerly intractable positions slotted into simplified conceptual pigeon holes. The present mood, represented in this collection, is to avoid confrontation or labelling for their own sakes, to attempt to transcend our mutual philosophical naïvety, and to ground our research in the real social problems that we face as geographers and as members of society.

Convergence also implies that we are moving beyond, or at least directly facing the dualism of structure and human agency which has replaced the geographer's previous quandary with the dualism of social process and spatial form, and which has formed the epistemological basis for dialogue between humanists and historical materialists. Convergence in this respect does not imply a choice of one or the other – human agency or structure – as the primary agent or motor of social change, but emphasizes the need for an incorporative understanding of their relationship and inter-dependency in the emergence of social history. At the same time two chapters, by Ley and Kobayashi, provide the basis for going beyond duality in resolving the structure/agency debate.

Diversity remains, and in fact dominates, this collection. The attempt to achieve common ground on the definitions of historical materialism and humanism, as a starting point, has been difficult, and it remains clear that at least some of the chapters in this collection are at odds in their use of terminology. This is a diversity within, however, rather than a diversity between two insulated and apparently homogeneous paradigms facing each other as self-contained opposites. This collection is not a dissolution, it is not a watering down, and it cannot be a search for common ground. It is rather an attempt to provoke, sometimes awkwardly, an opening, which allows us to unleash our imaginations in an attempt to learn not only from the newly seen (because newly created) reality around us, but from one another.

The search for common ground (Gould & Olsson 1982), as the recent spate of work on the history of geography indicates, is endemic to the field, whether motivated by a desire for intellectual security or simply by a desire for a shorthand with which to define ourselves relative to the academic community as a whole and to indicate our relevance and worthiness for funding. Apparently common ground, at least in its epistemological terms, is not available just now. What is available, and this collection makes this evident, is a community of concern, concern to understand, to communicate and to control the process of change. This particular outcome of the debate about structure and human agency comes from human beings steeped not only in an academic but in a political tradition which denies the separation of 'personal' and 'political', of 'what I do today' and 'what the world is', and which also denies us the luxury of discounting our own effectiveness. Believing that what you do in all aspects of your life, in your personal and your academic lives, matters, at once liberates and imposes heavy responsibilities. Very little is certain, almost nothing is finished, and the possibilities for creation are infinite.

As a consequence, this collection is diverse in a number of other ways. One of them is the differences in the modes of presentation, the ways in which different authors write. This is not only an indication of different points of departure, of the ways in which our different traditions have influenced us to write, but also an indication of unlearning common or standard forms of academic discourse. These chapters perhaps indicate that we have room for a diversity of styles, of ways of communicating. Above all they indicate that we have room, and we have need, for a variety of ways of expressing a new content. It is not just that existing concepts did not fit, and often obscured the content, but often that the words themselves did not fit, that the *given* did not fit. This, in some respects, is the most difficult kind of change. Often, even after we can see in new ways, even speak in new ways, we continue to write in ways which not only cannot contain, but which obscure, our seeing. Rigour is necessary, internal consistency is necessary, the kind of mutual respect which utilizes adequate footnotes and comprehensible language is necessary. But sameness is not only not necessary, it is not possible.

NOTES

1 'Restructuring' is a term which is coming to encompass a variety of social and economic processes, including 'de-industrialization', the growth of the 'informal economy', a more international economy and a greater reliance on home and community-based strategies. As the strategies of many people in the first world come to resemble more closely the strategies employed in third world countries, and the strategies of unemployed men to resemble more closely many aspects of the locality-based feminist tradition of organization, we would also expect some convergence in analysis of first and third world developments and a growing adoption of feminist concepts into pre-feminist analysis. Some indications of the nature of restructuring and its analytical implications are found in Gershuny (1978), Mattera (1985), Murgatroyd *et al.* (1985), Pahl (1984) and Redclift & Mingione (1985).

2 For discussion of the establishment of the positivist tradition in geography and the emerging break with it, see, for example, Holt-Jensen (1980, Chs 3 and 4); Johnston (1987, Chs 3, 4, 5 and 6).

3 Within the historical materialist tradition as a whole, this creativity resulted in a refocusing on the 'political' as in the works of Louis Althusser and Nicos Poulantzas, and on a re-examination of the basis of class constitution, in the 'French school' of urban sociology, most notably the works of Manuel Castells.

4 The re-examination of human nature and early discusssions of human agency went on in a variety of forums, including critical psychoanalysis (Brown 1973, Chesler 1972, Fromm 1962, Mitchell 1974), discussions of 'everyday life' within the historical materialist tradition (Bosquet 1977, Sennett & Cobb 1972) and of construction of life worlds in the re-emerging phenomenological tradition (Husserl 1970, Kockelmans 1967, Schutz 1962, Shmueli 1973, Tuan 1974). It was perhaps most forcefully articulated in radical and socialist feminist writing, including, for example, de Beauvoir (1961), Firestone (1971), Foreman

(1977), Greer (1971), Millett (1969), Morgan (1970), Rowbotham (1973, 1983) and Wandor (1972).

5 Some discussion in the early relevance debate included Anderson (1973), Berry (1972), Blowers (1972), Chisholm (1971), Prince (1971a), Smith (1971, 1973), and Zelinsky (1970). Some retrospective discussions of this period are found in Johnston (1987, Ch. 6) and Peet (1977).

6 Discussions of the origins and objectives of advocacy included Breitbart (1972), Bunge (1977), Corey (1972), Ernst *et al.* (1972), Roach & Rosas (1972) and Stephenson (1974). This work and concomitant developments also elicited discussion of new methodologies. See for example, Amaral & Wisner (1970), Harvey (1973, Part I) and Horvath (1970).

7 The most extensive critique in geography was Breitbart (1972). The seminal argument for a new theoretical basis was Harvey (1972a). A range of other proposals and counter-proposals is found in the comments on Harvey's article, included in the same issue of *Antipode*. Also see Slater (1977).

8 The concept of paradigm, if it was ever useful, is becoming less so. As Johnston points out (1983, 216) a paradigm model is popular because it offers 'a framework within which certain disciplinary changes can be accommodated (if not fitted) and it provides a propaganda base for the exhortation of further changes'. It is in serving these ends that the concepts of humanism and historical materialism have become so deeply embedded in geographical lore, and it is ideological clinging to one paradigm or the other that has resulted in their juxtaposition as mutually exclusive. At the same time, however, this does not become an argument for convergence for convergence's sake (Stoddart 1986).

9 See for example, Gregory (1981), Eyles (1981) and the exchanges among Ley (1980, 1982) and Greenberg & Walker (1982), and Duncan & Ley (1982) and Chouinard & Fincher (1983). Recently, the debate has shifted toward a more fine-tuned discussion of the specific relationship between agency and structure (e.g. Ley 1982, Gregory 1982).

10 Within the broadly historical materialist tradition, see for example, Dear & Scott (1981), Edel (1981), Gregory (1978, 1981a), Harvey (1975, 1978), Massey (1978), Sayer (1982b) and Smith (1979). Within humanism the emphasis has been on the social/cultural importance of the relationship between people and landscape (Gold & Burgess 1982, Ley & Samuels 1978, Meinig 1979, Relph 1981, Tuan 1979, 1982). Recent progress in landscape studies in cultural/humanistic geography is reviewed in a series of papers by David Ley (1981, 1983a, 1985a).

11 For discussion of the social engineering implications of positivism see Fay (1975).

12 This relational concept of the social whole in which the classic base–superstructure model and the 'articulated instances' of the structuralists are replaced with 'internal relations' is most clearly represented in the work of Bertrell Ollman (1976). These ideas have

been criticized and adapted into realist structuration by, for example, Keat & Urry (1975), Sayer (1982a) and Giddens (1976, 1977, 1979, 1981).

13 Husserl (1970), and those who adapted his ideas, concentrated initially on the structuring of individual consciousness (Collins 1968, Kockelmans 1967, Natanson 1963, Schutz 1962, Wild 1963). Within this tendency in the social sciences as a whole, humanistic geography developed initially as an attempt to incorporate the ideas and methods of phenomenology (for a review, see Jackson 1981). The idealist alternative is most forcefully argued by Guelke (1974).

A note on the sections

An understanding shared by contributors to this volume is the practical and epistemological indivisibility of theory, method and empirical research. Pratt, for example, argues that 'theoretically uninformed quantitative research is of little interest and of little use in explaining social process'. Foot, Rigby & Webber are explicitly concerned 'to show how abstract laws and concrete events may be linked to provide an evaluation of theory'. All the authors are concerned to understand how social processes have motivated new theoretical and methodological developments in geography and how we have responded to these developments both in terms of methods we employ and the empirical questions we ask. None the less, recognizing that it is not possible to say everything at once, the authors have chosen to emphasize particular aspects of these questions, addressing themselves primarily to the issues motivating their theoretical exploration and the empirical questions arising from this, or, alternatively, to the methods utilized to interpret social change.

The chapters in the first section of this book are therefore primarily concerned with the former questions, looking at empirical social changes and discussing these as motors of theoretical and methodological alterations. The chapters in the second section, although retaining this concern, are more explicitly concerned with methodological questions and the ways we might use humanism and historical materialism to address issues. Attempts to do historical materialism and humanism have required a rigorous examination of method, even a return to quantitative methods associated (often wrongly) with positivism. The questions of how we generalize dominate the chapters in the second section, examining the use of statistics as a mode of generalization and communication, the use of description as a mode of communicating, and the concept of landscape as one of the basic tenets around which the discipline is organized.

SECTION I
Issues

Introduction

The issues identified in this section represent the recent focusing of geographical concern on social problems, and show both how the premises of historical materialism and humanism have helped to focus concern, and why paradigms as such cannot be taken as cures for social ills. The issues presented here are not, of course, definitive of, or even salient among, the problems of modernity. We might, for example, have included chapters on environmental degradation, world peace, problems of global transportation and communication, or 'ethnic' and 'racial' marginalization, to name a few. Our aim here is not to be comprehensive, however, but to provide a number of examples of how geographical practitioners set research priorities, and mesh their theoretical perspectives with the exigencies of modern life.

Taken as a group, these chapters emphasize two points: (1) that ideology plays a major role in structuring both the empirical conditions of the world and the epistemological conditions according to which research is undertaken, and (2) that concern, commitment and relevance are not enough. The researcher must both maintain a spirit of criticism and be prepared constantly to adjust or adapt conceptual categories to emerging historical conditions.

John Bradbury sets the tone by focusing on the 'crisis of the 1980s' as having created a restructuring imperative within the contemporary international system of production. Although it is a truism to identify labour as the basis for human sustenance, it is not thereby simplistic to suggest that geographers be concerned with the provision of everyday needs at the most fundamental level. Bradbury shows not only that this level of activity is increasingly spatially extensive, but that there are specific problems of spatial discontinuity associated with the recent recession and the possibility of its recovery. The geographic implications of restructuring are evident in cycles of renewal and deindustrialization, and concomitant investment and disinvestment. Further, these conditions bear upon questions concerning technological change, changes in modes of production, the role of the state and the international division of labour.

The issue of integration is continued by Mackenzie, who argues for a simultaneous reconceptualization of empirical and theoretical categories both to include gender as a necessary category in the constitution of social life, and to recognize that history involves the simultaneous alteration of gender and environment. Her claim that 'the link between an androgynous human and the environment is the active use of space and time by women and men' is based on the dual roles of women in the productive and reproductive spheres being both constituted and transcended through active, political agency.

Jeanne Wolfe focuses on the relationship between the academic and the practitioner, to show that theory really *does* matter, and that the landscape is a direct (although not therefore perfect or uncomplicated) result of theoretical positions which inform policy makers. This statement emphasizes the collective responsibility of theory makers to consider the vast array of linkages between theory and environment, and to recognize that the importance of theory rests not only in its empirical testability, but also in its normative, ideological underpinnings. The city we inhabit is an extension of ideas that have a direct influence on people's lives.

Richard Harris provides a further example of people redefining their social roles to achieve political, and thereby material, ends. This chapter emphasizes the need to consider real, historical actors within specific, even particularistic, contexts even if (as Marx tells us) the efficacy of those individuals is very limited. Like Mackenzie, he emphasizes the simultaneous constitution and alteration of human and environmental relations.

1

The social and economic imperatives of restructuring: a geographic perspective

JOHN BRADBURY

INTRODUCTION

This chapter will present a means of understanding the medium and the method of economic and social restructuring from a geographic perspective. The term 'restructuring' is used here to describe the empirical experiences during a crisis event or a phase of adaptation and change in a whole system of production, be it at the local, regional or national level. As we use it here it embraces the modifications in the production system itself which have repercussions on the location and distribution of work and industry and on the viability of the whole process of capital accumulation. More specifically the term suggests both positive and negative processes: positive restructuring when new production relations are generated and negative during a phase of deindustrial-ization. Spatially these differences have some obvious repercussions, to wit: decline in one region's industrial base may be matched by an equivalent growth in an adjacent zone or even in another country (cf. Mahon 1984, Thrift 1986, van Ameringen 1985).

The theoretical concepts explored here are derived from observations of empirical events in the recession of the late 1970s and early 1980s. In this period it became clear that a new perspective was needed to see the processes of restructuring both in terms of their context holistically and globally while retaining an important sense of the internal complexities (concretely and regionally) and the implications for capital, workers and the state. The recession pointed out quite clearly that swift movements by firms in the developed, developing and underdeveloped worlds were in response to declines in profitability in some locations and the growth and relocation of investment in others. These movements resulted in the decline of industry in some settlements and regions and its growth in others. The changes in locations of firms were matched by some equally significant reformulations of the work people do, especially the mix of skills, the introduction of new technology to replace labour, the search for pools of cheap and pliant labour in 'greenfield sites' and the extended deployment of youth and female workers in factory sites and other places of work around the world. All these changes took place at an increasing rate during the recession as capital, generally with the compliance of the state, embarked on a period of intensive restructuring.

The theoretical parameters and the social and economic consequences of these changes are the subject of discussion here. The chapter is divided into four parts. In the first section we outline the framework of analysis and discuss some of the causes of the current round of geographical and economic restructuring. In the next section we address a specifically materialist view of restructuring drawing on current literature and observations of empirical trends. In the following section we focus on what happens to labour in a restructuring event; and in the final part we look at a comparison of first and third world restructuring events.

WHAT ARE THE CAUSES OF GEOGRAPHICAL AND ECONOMIC RESTRUCTURING?

In capitalist-based societies the fundamental determinant of the restructuring of geographic space and of the social and economic relations of production is the arena of production itself. The pattern of motion of investment or disinvestment of capital imparts its own particular features on a regional economy. The spatial component is important because the production system itself, having become increasingly internationalized, operates within and between regions and across national boundaries. As such, the process of restructuring has taken on an international character with regions, towns and firms vying for favourable conditions and policies. Indeed, as Palloix has argued, the successful large-scale expansion of multinational firms can no longer be accomplished within a single region or nation state and must be based on an ever-widening circle of industrial investments and locations (Palloix 1975). This too makes the restructuring imperative increasingly complex as large firms are able to expand and contract their systems of production and sets of production relations in accordance with differential spatial rates of profit.

The restructuring process embraces investment as well as disinvestment (see Table 1.1) which in turn presupposes crises – and crises demand solutions. It has been argued that the two processes must be applied simultaneously in order to maintain viable conditions for capital accumulation. In the international crisis of the late 1970s and early 1980s, however, the opportunities for switching investment from one industry to another or even between sectors were limited. These limits embraced energy costs, technological traps, high relocation costs, the increasing pace of technological change and the immobility of fixed capital, especially in some of the traditional 'sunset industry' regions such as in the northeastern United States and in the north-east of England. Thus the changes in the amount of capital required to be invested in machines, technology and fixed capital in general increased exponentially, placing considerable stress on investments and on investing institutions. The result was a phase of increased disinvestment in certain sectors and areas and the growth of new industries in other locations.

We must caution against seeking the causes of restructuring entirely within a local area, region or even within the nation state. Actual changes are most likely to be the result of interaction between internal and external components of the international division of labour. This is particularly so within relations between production units such as factories, branch plants or foreign-owned firms. Clearly the developments in the productive forces embracing capital and ownership are dominant. But they exist in a conflictual and often contradictory relationship, that between the workforce and the social and technical division of labour. The dynamics of restructuring must therefore embrace an understanding of the rate at which these relationships develop and change; conflict is an integral part of the relationship especially in the workplace, but as has become obvious in a number of instances, some of the immediate conflict has been removed by the employment of non-union workers.

We do not subscribe to a linear framework of explanation of restructuring, but rather to a cyclical and crisis-oriented form of change within different modes of production. Linear models suggest that restructuring and modernization are directly associated with the process of growth. Needless to say this view is not always valid, because evolutionary change is not a matter of steady moves from one stage to the next. Although there is some evidence to suggest that plateaux of development do actually exist it is difficult to provide supporting evidence of the causes of entrance or exit from each 'stage' of development. Nor do such models allow for ready analysis of the processes of stagnation, decline, restructuring or uneven development which are an integral part of the process of development, at least within the capitalist mode of production (N. Smith 1984).

To this general schema of events we must add the comments made by Holloway & Picciotto (1978). They suggest that the process of restructuring involves three main areas: an increase in the rate of exploitation; devaluation and destruction of constant and outmoded fixed capital; and the redistribution of total surplus value through state policy from less productive state expenditures and inefficient capital. It is also apparent that the tendency towards a falling rate of profit (see Table 1.1) and the struggle between capital and labour and between different classes of competing capitalists provides another arena in which conflict and restructuring merge. The conflict evident in these situations appears during periods involving job change, the redivision of labour and the restructuring of skills.

We can also turn to a framework which permits the analysis of the internal and external conditions of restructuring, within a mode of production at specific times. In this manner, it is possible to analyse both the dialectics of change and the apparent contradictions which exist within the different levels of development. Furthermore, the impetus for change is evident within cyclical patterns of upswing and downswing, although the occurrence of cyclical behaviour in and of itself is not an adequate explanation of economic change and restructuring. Indeed, the engine of change and the motive force of the mode of production hold the reasons for the occurrence of uneven development and the need for constant restructuring.

Although there is some general recognition of the problems of linear and cyclical development, there is one internal debate which encompasses all these issues: that the processes leading to uneven development are a necessary prerequisite to continued capital accumulation; or that the spatial and sectoral manifestations of unevenness are the logical outcome of different phases in the overall restructuring process. The answer to this question lies not in following one or the other, for both in their own ways are inappropriate. What may be more profitably pursued are the actual premises of restructuring and the underlying logic of the mode of production being examined. What will become clear in the course of this chapter is that changes in the mode of production, at the level of mine, mill and factory, induce and effect the restructuring of capital and the social and economic relations of people within and outside the workplace.

A MATERIALIST VIEW OF RESTRUCTURING

A materialist view of restructuring embraces those conditions within the mode of production which either impinge on each other or interact within the constant ebb and flow of investment, industrial production, employment and cyclical behaviour. Much of this type of change can be understood under the rubric of a crisis theory of economic behaviour in which restructuring is both an end and a means to continued capital accumulation (Harvey 1982, Wright 1978). The crises themselves may be derived from any number of sources, including fluctuations in both supply and demand of inputs and outputs. These I have noted as 'trigger events' in Table 1.1. For the most part these events embrace interruptions in the smooth running of a production system, whether they be from the state, capital or labour. These 'trigger events' are followed by 'general process changes' embracing the reformulation of the production relations in a plant or a community. These events in turn have a 'structural outcome' which has been summarized by Massey & Meegan (1982) as technical change, rationalization and intensification. The final geographical impact is a result of the combined effects of trigger mechanisms, general process changes and structural outcomes.

As Table 1.1 shows, we must consider various forms of capital and credit shortages and sources of finances when looking for explanations and causes of restructuring. For without capital the production system itself is unable to continue. Furthermore, as is notable in the crisis of the early 1980s, there is an urgent search for new technology which will replace labour. This technology is being produced and reproduced by companies involved in the production of technology *per se* and in the overall production system itself. In some instances old machinery in 'sunset' industries wears out and is replaced entirely by new technology; in other cases, new parts and new machines replace the older machines at the same, or perhaps a slightly more sophisticated, level of technology.

Technological change *per se* is one way in which restructuring

Table 1.1 General processes of geographical restructuring

Trigger events	General process changes	Structural outcome	Geographical impact
Cyclical downswing	Restructuring of corporate investment plans in	Winding down/up	Community – rupture of work/social networks, temporary or permanent unemployment
Profit decline – absolute and by rate	machinery and repairs – constant, partial or complete reproduction of fixed capital	Plant openings/closures Disinvestment/investment Machinery movements/ acquisitions	– layoffs
Production bottlenecks		New inter–industry linkages	– migration
			– new job cycles
Decline in capacity and quality of fixed capital	Labour process changes – modifications in rate and intensity of work	Intensification – prolonged/temporary	Plant – decline in quality of infrastructure
Technological peaks and plateaux in plant/ machinery efficiency	Job skill changes derived from new technology	Rationalization – continuous/periodic	– changes in local and regional industrial linkages
Class conflict in the workplace involving capital and labour:	Reformulation of class positions and conflict regulation	Technical changes – maximum/minimum	– location/relocation of plant in 'brownfield' and 'greenfield' sites

– Intercapitalist conflict and competitive behaviour	Products aligned toward consumer needs	State intervention – tariffs – protectionism – financial aid	Restructuring of markets – nationally and internationally
Overproduction or underconsumption?	Search for lower costs of production: via energy efficiency (heat), machinery changes and innovations	– market allocation – labour regulation – state capital/valorization – legitimation of capital's actions – import and export controls – non-tariff mechanisms – anti-dumping laws – quotas	Changes in intensity usage in different regions
Commodity and product cycle crises			New national production systems in developing world – sunset and sunrise industry syndrome
Raw materials and energy costs	State action: deployment of regulatory mechanisms and state capital		

can take place. We must be careful, however, to distinguish between technology which 'leads' a major phase of restructuring, and technology which is implemented by lagging firms to catch up or to maintain levels of productivity and competitiveness. In both cases the technology may be considered as an innovation in that it adds to or substitutes for an older form of technology; and whatever its status it will probably have an impact on the numbers of employees and their skills and job tenure. Technological change has a number of effects whether it is in a leader or laggard status. The new technology may introduce plant-saving techniques making it more efficient, at least in the short term, which will mean that job levels may be retained. The longer-term incremental substitution of labour with capital (workers with machines and nowadays 'hi-tech') can result in a slowing down of new employment and ultimately in the absolute replacement of labour (cf. Thwaites & Oakey 1985).

New technology is part of the system of machines in a production process. Given the ultimate aim of capital to substitute for labour wherever it is financially possible, the place of technology is very important. Technology is therefore a major variant in the restructuring agenda both during normal phases of production and during periods of unusual growth or crises and downswings. This is one good reason why technological innovation is constantly sought, by firms who can afford it, in order to improve the efficiency of existing fixed capital. The addition of sometimes simple devices may mean that machines can be speeded up or permitted to adapt to other or additional tasks. This type of innovation can occur constantly, and in some instances it is required in order to maintain the integrity of older or 'mature' production machines.

The competition for innovations and the speed at which they can be commanded and brought into production bears on the relative success of an industry's restructuring profile. Competition and secrecy within the field of innovations are often high, as is the cost of maintaining a constant vigilance in the 'technology race'. Those firms and even whole sectors which do not maintain spending in replacements of machines and in new technology may not maintain access to new technology and therefore may not survive a restructuring crisis. Occasionally, they may survive by 'Heath-Robinson' or 'wire and string' modifications to existing plant, but for the most part their access to industry-wide technology is limited by their lack of capital or by a lack of 'modern management' at the plant (Malecki 1986).

Competitive behaviour both within and between sectors also

provides the impetus for restructuring. The level of competition *per se* is dependent on the success of the production system and the capacity of firms to create a production unit and a set of labour/capital relationships which can match or overcome those set up by other firms. We must also note, however, that not all firms compete at all times; there is plenty of evidence to suggest that different levels of co-operation do exist at the level of sharing information, joint ventures and interlocking directorships.

The crisis of the 1980s demonstrated a number of significant strategies in management and labour relationships in restructuring. In many instances the firms under stress opted for a reduction and 'rationalization' of their workforce. In most cases, this meant that labour was substituted for by machines of increasing sophistication and productive capacity. Where no technology was introduced, firms endeavoured to cope by reducing the numbers of workers on the job, increasing the hours worked and the intensity of the work process. Restructuring of this nature was notable in many large- and small-scale industries in Canada and elsewhere.

Some firms chose to conduct their restructuring as a series of combined events during the recession, suggesting that the weakened position of labour and the need for technical innovations in a downswing were sufficient to trigger a series of labour rationalizations and the introduction of new and sophisticated machinery.

In terms of the schema of events noted in Table 1.1 the geographical impact of the changes becomes evident at the level of the plant, and of the community where workers lost their jobs and were required to undergo retraining, relocation (migration) or a period of unemployment. What became clear in the restructuring of the 1980s was that many workers would no longer be employed in the industry they received training in and perhaps had worked in for many years. Furthermore, many of these workers would be required to migrate to other communities in search of work. Lastly, many more would not be employed again, especially those of a particular age group approaching retirement and even those in middle- and late-middle-aged categories. In some industries workers with long-term standing were retained instead of short-term workers; such was the case in a number of plants in the Canadian steel industry in the 1980s (Bradbury 1985).

Although the sources of crises noted in Table 1.1 are readily observable, we must note that each may interact with the other or in turn respond as a countervailing force if corrective measures are used. The prevailing condition, however, is one of crisis behaviour and modification both within and between economic sectors. In the event of apparently traumatic events, the state is generally called on to intervene or to promote or modify the direction of change in the restructuring process. The state is an integral element of the system itself and therefore of the restructuring events which occur. The state may become involved at the level of policy, ideology, action in the market place and in the production system itself. In all events it has become apparent that the role of the state is changing constantly, depending on the direction of the needs of the restructuring process invoked by capital and on the ideology and degree of involvement of the state in the production process. Whichever position is adopted by the state it is also evident that it responds to the call of the private sector in times of need either selectively to apply state resources and finances or to shape policy and control and protect markets and production relations. In the event of the rupture and closure of a plant or indeed of a whole production system, as was evident in the mining industry in northern Quebec and Labrador in Canada, the state may also be called on to supply financial support for the companies and the communities which were closed down (Bradbury 1984, 1985, Bradbury & St Martin 1983).

What causes the mode of production to change? Answers to this lie within a multi-causal set of relationships surrounding the drive toward constant capital accumulation. To this we must add the profit imperative and the constant climate of competition within the overall centralization and concentration of capital. Similarly our analysis must embrace consideration of the effects of contractions as well as expansions in the system of production. For no system expands at an obvious or continuous rate, and all are fraught with problems of slow growth, recessions and downswings. Furthermore we must also allow for the effects of changes in the rate of profit within each sector and even within each enterprise; all of which will dramatically influence the capacity of each unit of production to continue investments in both fixed and variable capital, in machines, labour, people and social infrastructure.

It is apparent in the real world of economic and social geography that there are many factors which militate against a simple linear growth pattern and are rather faced with a pattern of continuous expansion and contraction processes. Indeed, as we shall see, these are built into the very logic of the system of production. The dialectics of expansion and decline, renewal and deindustrialization, employment and unemployment, all under the rubric of restructuring processes, are the logical outcome and consequences of the imperative of capitalist growth. That this takes place unevenly over space, time and by sector is not an unusual or unpredictable phenomenon. Indeed it is becoming increasingly obvious that the cumulative effects on people and places are evident in the waves and cycles of upswing and downswing. The internal dynamics of these waves, while they do not deploy a driving mechanism of their own, are evidently influential on the rate, location and sectoral behaviour of upswing and downswing (Gordon *et al.* 1982). Thus the dynamics of the production system itself provide both the medium, cause and need for restructuring. And, although the need for restructuring is constant, it occurs at different levels and sectors in different locations. We must therefore caution against viewing the current phase of regional and industrial restructuring as something new or unique. Rather, it is the acting out, albeit unevenly, of the cyclical dynamics of the system of production itself.

There is some evidence to support the 'mode and submode' construct, but the scale of change required to invoke this type of explanation may sometimes exclude non-conforming or non-fit restructuring which is implemented to iron out or ease small fluctuations in the production system (Gibson & Horvath, 1983a, b).

Gibson & Horvath's case of transition between submodes is illustrated by major alterations in technology, the social division of labour, changes in the work process, confrontation between the banking sector, industry and commerce; and the reshaping of relations between the state and productive capital. They also point out that major forms of change in a mode of production result in institutional and ideological changes which support the general direction in which the restructuring is moving. In their framework, economic and cyclical crises, originally outlined by Wright in 1978, form the trigger mechanisms, and the medium for restructuring (Wright 1978, cf. Gordon *et al.* 1982).

According to Massey, firms involved in a crisis situation must escape the outcome by restructuring. Massey argues that several options are opened up during the crisis phase and the subsequent shakedown which takes place in restructuring. She summarizes the overall process into three summary types, to wit: rationalization, intensification, investment in technical change. Massey & Meegan provide evidence of this form of change in the United Kingdom, but we must be sure of the evidence for such changes elsewhere. It is clear that research and case studies of restructuring and its social and economic impacts are needed in other countries, both in the first and third worlds to complement this work (Armstrong & McGee 1985, Massey 1984, Massey & Meegan 1982).

WHAT HAPPENS TO LABOUR IN A RESTRUCTURING EVENT?

The general processes of restructuring can be researched at the level of the plant, sector and region (Grayson 1985). At the first level, in the workplace, we must observe changes in the labour component of production. Are there changes in numbers, gender of the workers, levels of skills and hours worked? It is apparent that within separate workplaces and work processes the shaping and texture of restructuring takes different forms. What took place in the 1980s' recession was a reshuffling of the balance of capital and labour within the specific format of fixed and variable capital. The strategies deployed in the drive toward restructuring suggest that separate firms are placed in a position of revamping their production methods or lines *per se*, or of decreasing the labour content of the commodities they produce.

The reduction in the number of workers is the most commonly deployed method and first step in rationalizing a labour force at the level of the plant. The second step is to rationalize the numbers of hours worked with fewer workers. Changes of job hours and the revamping of shift structures can reward industry with similar production rates and lower labour costs. At the same time as these labour reductions occur firms have been known to replace machinery and to reorganize the speed and output of machinery in the plant. Firms may also 'furlough' parts of their operations, especially where there are duplicates of machines and processes or multiple assembly lines. In such cases the reduction of labour can go hand in hand with partial plant closures. All of these methods go under the general name of rationalization of both plant and labour in the restructuring process.

The absolute reduction of labour *per se*, however, is only one of a number of such strategies. Workers may be moved from place to place, from city to city and across regional boundaries. Workers may be reallocated to different work stations or to different plants in an attempt to mobilize a lower number of workers around a reduced number of work stations. There is also evidence to suggest that computerization of the spatial allocation of workers within plants and work stations has led to a higher rate of movements within a plant and to a redifferentiation of skill structures and job types. Instead of workers becoming specialized within one particular area of a plant or some other work area such as a hospital we see an increase in the number of tasks expected of a worker and a consequent reduction in the area of specialization.

It seems we are now entering an era of 'reverse-Taylorism' in which the workforce will be required to have multiple skills of perhaps a limited type but nevertheless expected to cover a wider range of jobs. Labour reallocation in both a spatial and technical sense is, however, somewhat more difficult in some sectors, especially those which over the years have built specific skill structures and work stations into the fixed capital of a plant. Similarly the adaptations are limited in circumstances where workers cannot move to new stations and use new skills without expensive retraining.

For the most part the major moves toward labour restructuring in the recession of the 1980s resulted in absolute job loss in some sectors and gains in others, although there is no evidence to suggest that these changes were equivalent or mutual. Though some have argued that absolute numbers are likely to be replaced after a job loss it is rare that such claims can be verified and even more difficult to find jobs in the same area, sector or with the same skills. Most movements require costly retraining to absorb workers into the same plant or industry; debates ensue about whether the state or the private sector should pay. Similarly the job skills are seldom equivalent in different sectors and the job loss and job gain statistics used to date cannot differentiate between the types of skills that are mobile and those that are incompatible. It is, for instance, difficult to imagine or to persuade miners and steel workers to become shopkeepers or nurses.

There are some job types, it seems, that do not match the technical, social or cultural dispositions of workers in some localities. Nevertheless in some countries such transfers are touted as the answer to the wave of displacements created by disinvestment and restructuring in the 1980s. It is argued that the fabled 'post-industrial society' will enable this form of mobility and in fact will demand it. At present, however, there is no foreseeable guarantee that jobs in the service sector have their equivalent in terms of locations, numbers, salaries, benefits and skill profiles. Smooth movements or transfers of labour in a cross-sectoral restructuring activity are thus very difficult to achieve satisfactorily. If moves do occur after a plant closure then workers tend to orientate toward lower-skilled areas, the service sector or to periods of unemployment. Such was the case with displaced steel workers between 1978 and 1983 in Canada (data from Employment and Immigration Canada).

What Harvey (1982) has referred to as a 'spatial fix' in terms of fixed capital in machines and factories, has its equivalent in communities and workers too, especially in zones of older industries where several generations of families attached to industrial communities predominate. Movements of firms away from such areas or the revamping of job places, skill requirements and work stations can act with equivalent trauma on the labour force. What is most common in such areas is the restructuring of capital followed by revamping of both conditions, location and remuneration of work. This is most noticeable in regions and sectors attached to sunset industries where firms become detached from older sites and in this footloose state either relocate, revalorize their capital into new forms of production or lose the absolute value of investments through bankruptcy. In such regions labour is forced to become mobile, if this is possible, or it becomes dependent on the state or on some new form of diversified economic activity, often based on the service sector, or on new industries which may locate in order to obtain access to a cheap, by then cheaper, labour force.

Increasingly, labour–management relations and class conflict in the community and the workplace are seen as factors which must be considered in the process of geographical restructuring. Evidence drawn from the Canadian experience in the 1980s would suggest that the major restructuring to come out of free trade with the United States may bring about job losses in some areas and job gains in others; the balance of such moves will depend on the types and locations of industries affected in Canada.

Trade unions are concerned about the protection of jobs and the maintenance of wage levels and standards of living. In periods of upswing, wages are increased through bargaining, but in downswings negotiations are marked by intense bargaining and the withdrawal of labour as well as the denial of the right to work in the factory or plant (Bradbury 1985). The conflict inherent in such situations becomes focused on control of the work process and control over the workplace. Firms may parallel this conflictual activity with a period of contracting out of certain parts of a production process which may exacerbate the conflict or remove parts of the local labour force entirely.

Conflict also arises when layoffs and dismissals occur in a plant and in a community. These bring about displacements and destruction of community linkages and structures. There is evidence in North America and Great Britain that the recession of the 1980s induced considerable restructuring of this form in some of the older 'sunset' regions embracing older industrial systems and primary production plants such as steel mills, mining towns, textiles and clothing mills. Such was the intensity of this conflict that tensions focused entirely on the attempts to retain jobs and maintain the integrity of local operations, communities and working-class life and culture. Attempts were made to negotiate with factory owners to keep jobs or to trade off with community facilities or state compensation. Workers in some areas, especially in mill towns related to steel or textiles, even went so far as to attempt to purchase their local plants and production facilities. A number of strikes took place before and during the closures in an attempt to pre-empt or prevent the closures of the plant.

The restructuring which takes place in the economic arena has direct repercussions in the workforce and on the social relations which exist among workers in the workplace and in the community. The workforce which is 'released' through 'rationalization' and restructuring and the subsequent relocation of jobs and workers has been increased not only by the proletarianization of workers *per se* but also by increasing the numbers of women and youths in the labour market. This particular restructuring strategy enables business to lower the average wage and change the location of work both within the city and from region to region; it can also increase the intensification rate and increase the number of workers per household. This 'new' labour, consisting of women and less politically powerful groups such as immigrants located in 'ethnic' areas of the city, is often paid very low wages. Married women, for instance, are paid low wages because it is sometimes argued that they are either second earners or primarily dependent on some other wage earner. The same strategy is also deployed in third world countries where women are paid at significantly lower rates and employed for very long hours. The oft-quoted example of female electronic component assembly workers working long hours at eye-taxing and stressful work and then fired after four to five years because their eyesight has failed, is an indication of the kinds of redeployment of industry and labour which came out of the international restructuring of capital and labour in the post-1970 period. We must also note that the low-wage female-dominated workforce is also likely to be non-unionized and credited with low levels of political awareness and strength. Furthermore, they are located within regional labour markets which give ready access to a constant supply of migrants and an ever-changing cohort of compliant and youthful workers (Froebel *et al.* 1980).

In the next section we discuss aspects of the movements of capital to the third world and point out the linkages between the systems of industrialization. We note that the 'global process' is indeed a result of the continuing industrialization of capital and an extension of a network of production units, albeit into different cultural milieux at widely dispersed locations. The overall linkage, however, courses through the network of finances, ownership and control of international capital, and the search for new pools of labour as well as new markets.

RESTRUCTURING IN THE THIRD WORLD

Although much of the literature on restructuring has been written with a first world perspective, there is still a great deal of restructuring in the third world, especially as a result of the long-term internationalization of capital and the latter day movements of some specialized manufacturing and assembly firms to so-called 'off-shore platforms' and 'labour havens'. The concept of the 'global factory' means that we must organize our framework of analysis to include both the moves of first world firms to the third world to engage cheaper labour, and the subsequent impact on workers everywhere of these moves.

The form of multinational intervention in developing countries has changed significantly in response to a new relationship between national and international capital, with the latter preferring to leave the total ownership of all production stages in favour of different forms of joint-ownership and concentration on the most profitable arenas. By means of the multiplication of joint-ventures, production licences, subcontracting and the moving of entire production units to developing countries, multinational firms have embraced both deepening and broadening of capital.

The choice of restructuring strategies in the developing countries depends on the particular nature of the labour market. However, the persistence of non-capitalist or 'traditional' social formations mediates the subordination of labour, even where the labour form is present (but not universal). Thus third world proletarianization is not reduced to the availability of the workforce recently excluded from agriculture. There is as well a persistent form of subsistence economy which embraces a 'domestic community' or 'family household' in the modern capitalist sector in which there is no 'free labour'. The persistence of this subsistence economy in both the informal and the formal sectors affects the supply of labour and the nature of the proletarianization taking place. The result may be found in the various forms of semi-proletarianization which exist in many developing countries. Here we note that the presence of this non-capitalist sector allows restructuring to reduce wages by increasing the supply of labour of 'willing workers', through the increased participation of women and youths, or by including workers displaced from a previous mode of production. Much of this workforce may be made available only through the changing commitment of the family division of labour. It allows industry to lower wages while permitting the intensification of activities and increasing the numbers of workers per household.

In developing countries the moves of high-technology firms in the 1970s and 1980s are the most obvious and dramatic wave of a transfer of productive investment to off-shore labour havens and to special economic zones and free-trade zones. For here the wages are low, benefits to workers few and taxation breaks favour those firms most suited to using the 'long and nimble fingers' of workers. Many cases have been documented of the moves of electronics and other factories to the third world labour havens and, while these form part of the concept of the global factory, they must be distinguished from the forms of labour restructuring evident in local industries adjacent to these enclave industries (Froebel *et al.* 1980). Here the evidence differs on the impact of indigenous industrial growth versus that derived from 'off-shore' foreign investment. Some argue that the spin-off effect of these foreign investments is high whereas others argue the opposite, suggesting in fact that they operate as foreign enclaves. On the other hand it has also been argued that job creation rates in local areas of industrialization in some parts of the third world are much higher than first thought. Between 1960 and 1979, for instance, industrial employment in less developed countries grew at a rate twice that of some Western countries between 1900 and 1920. The effects of this relative growth in the present conjuncture have, however, been masked by expansion of the overall growth of population (Squire 1979).

The most powerful tool of restructuring in the third world in the post-1973 period was the move to 'labour havens' which occurred in conjunction with the start of a global downswing. This was the start of a long-term trend toward switching the location and the sectoral operations of numerous multinational corporations whose traditional arena was the first world (McGee 1984, p.9). In this phase a spatial movement of capital took place in response to the downswing crisis of international capital; this was precipitated by a 'switching crisis' which involved the movement of capital as one response to the problem of falling rates of profit (Froebel *et al.* 1980).

In developing countries extensive use has been made of new male and female skilled and unskilled labour. More has occurred since multinational companies have become involved in the transfer of production systems into the formal industrial system of some third world cities (Armstrong & McGee 1985). In terms of gender differences in the labour force it is clear that the contribution of women, be it in the formal or the 'informal' sector, has become increasingly important. Within the phases of general restructuring, however, as in each local region, women face a host of obstacles, including: inappropriate education and training; lack of capital; resistance by institutions to grant loans and credit; and limited employment and income opportunities. In those cases where jobs have been created in the enclave sectors, say in electronics, the work periods are long and hard and as a result the burnout rate and turnover are very high.

Although the number of women working in Asian industry has more than doubled over the last two decades, they are mostly concentrated in a narrow range of low income occupations, with inferior terms and conditions of work and little opportunity for advancement (UNIDO 1984). Furthermore, their cultural status and position limit their access to jobs in the first instance and predispose them to quicker exit from the workplace during restructuring crises which include reductions in the numbers of workers. Similarly the demand for their cheap labour has been accompanied by sexual discrimination and damage to health in unprotected situations. Nowadays there is widespread concern at the conditions in workplaces where jobs are boring, repetitive, dehumanizing and in some cases dangerous to physical and mental health (UNIDO 1984, p.25).

Similarly, in a number of developing countries there are strong cultural prejudices which influence the manner in which women are incorporated into the labour process in the current phase of international industrial restructuring. They have fixed and subordinate roles in the labour market, which means they are designated positions within the international division of labour – on the very bottom rung. Women's status is traditionally low and this simply reinforces their position in the social division of labour in the new industries. At the same time, however, many urban dwellers, and now some rural women, have found jobs in large-scale assembly and components factories, but mostly under unfavourable working conditions and incomes. Such has been the impact of job changes that although women have obtained wage incomes and earnings these are often in addition to their traditional domestic responsibilities.

CONCLUSION

Our understanding of the geographical implications of restructuring has drawn on a number of issues surrounding the events in the recession of the 1980s. We have examined some of the broad trends in both first and third world development which embrace the processes of restructuring. It is clear that restructuring is both an economic and a social phenomenon, which embraces the production systems of both worlds. Increasingly, we see that the efforts to contain the crisis have expanded into the third world as the internationalization of capital has proceeded apace. Where firms have moved to the third world they have tended to employ cheap labour and the subsequent industrialization has followed the path of enclave development.

Restructuring follows some well-defined strategies, namely the rationalization of labour, capital and the production system. The geographical implications of this process must be appreciated in both upswing and downswing periods. For these periods are ones in which change takes place very quickly and in which companies, governments, workers and communities are all subjected to extraordinary pressures. In the downswing phase, as was experienced most recently in the early 1980s, we noted an acceleration of movements of firms and capital between regions and between sectors. What this meant was the redressing of older regional systems and the generation of new sets of relationships. What is most significant for us, and what urgently requires new fieldwork and research, are the changes which have occurred and are probably still occurring in those settlements and communities of workers undergoing restructuring. Such settlements as those which embrace the older 'sunset' and newer 'sunrise' industries must be looked at very carefully in order to understand the type of change which is taking place and the implications for the material wellbeing of the people who live there.

2

Restructuring the relations of work and life: women as environmental actors, feminism as geographic analysis

SUZANNE MACKENZIE

INTRODUCTION

For some years now, I have been meeting women with very complicated schedules. These are women who get up at 6.30 in the morning, dress and feed their children, drop the older children at school, drop their husbands at work, drive two miles to babysitters to drop their younger children, pick up co-workers and finish their journeys to work by driving four kilometres to their wage jobs. At lunch they shop for dinner. At four they drive home – they work part time. On their journey from work, they make journeys to shop, journeys to pick up younger children from the sitters', journeys to drop off their co-workers, journeys to pick up older children from friends' homes. When they get home, they make dinner, talk to the children, put on a load of laundry, try to talk to their husbands over dinner, sweep the kitchen, fold the clothes, all in their leisure time.[1] Some women have busier lives because they have two money-earning jobs, one at night or at home. Some have no car, no automatic washer, no husband, or they have disabled children. In those often unquiet moments when I talked to them, almost all wondered out loud how things turned out this way, how other women coped, and what would happen if one of the apparently infinite number of variables they were juggling in their lives was altered.

It is lives like these and questions like this which motivate and which form the empirical and political content of contemporary feminism. Within the last decade, feminism, as politics and analysis, has colonized and challenged virtually all disciplines in the social sciences and humanities.

The emergence and extension of feminist research in academic discourse has been a complex and often insurrectional process. But it has generally followed a pattern. In their initial work, women and men, inspired by changes in women's patterns of activity and by an increasingly commodious and analytically sophisticated political feminism, concentrated on exploring and documenting women's historical and contemporary roles. They amassed empirical evidence which made women visible as historical and social actors. But about the time that research on women became an accepted – or at least unremarkable – part of at least some disciplinary agendas, researchers began to realize that empirical documentation was not enough. Most of this pioneer work had been utilizing existing disciplinary frameworks to examine this new content. But doing so almost inevitably involved squeezing the fluid lives of women into often inappropriate categories; subjecting the 'petty', 'mundane' details of maintaining human life in capitalist society to the scrutiny of formal, scientific perspectives. And there was often a mismatch between this new, mercurial content and the existing frameworks of academic discourse. This caused researchers to raise a series of questions about the adequacy of the theoretical and methodological underpinnings of their disciplines, and to look for ways of modifying these underpinnings to accommodate the new questions they were asking. It became evident that neither changes in women's roles nor feminist analysis were phenomena which could be dealt with by adding a new empirical subject of study, or by stretching existing frameworks. These changes and this analysis required a fundamental questioning of the basic concepts and methods of disciplines, and the development of new concepts and methods. It was this realization which marked the transition from research 'on women' to feminist research – work where the content and political priorities were seen to require an independent theoretical and methodological base.

Similar kinds of questioning, with similar creativity, were being carried out by non-feminist researchers, and were leading to an increasing interest in other non-positivist perspectives, notably humanism and historical materialism. These entered disciplinary discourse at about the same time as feminism, largely motivated by the same social changes which underlay women's altered roles. And whereas the specific form and often the content of the questions differed from the feminist questions, researchers who asked about social conflict, about class relations and about social meaning and human action experienced a similar process of emerging dissatisfaction with the capacity of the given disciplinary frameworks to comprehend their questions and the changes which inspired them.

These three perspectives – feminism, humanism and historical materialism – are thus related both through their history in Anglo-North American social science and through a set of questions which not only arise out of the same social processes, but which are largely incomprehensible within positivist frameworks. And, within the last few years, as all three perspectives pass through their pioneering stages, their common concerns are becoming clearer.

A three-stage process of empirical exploration, theoretical creativity and convergence with other theoretical frameworks can be discerned in the growing body of feminist-inspired research in geography. Having developed a substantial body of empirical work on women's historical and empirical environmental relations, geographers began to examine the implications of this research for the discipline as a whole. This first, empirical stage could fairly be called work on the 'geography of women', the second, emerging theoretical stage, 'feminist geography'.[2] It also appears that the parallels between feminist geography and some recent developments in humanist and historical materialist work in the discipline are becoming evident, primarily focused around our conceptualization of the relations between people and the environment, or the recasting of the 'social process and spatial form' debate into the 'human agency and structure' debate. In fact, these three perspectives have, in combination, been instrumental in the re-emergence of the view of geography as concerned with human environmental relations as an alternative to, or in opposition to, 'spatial science'. And although feminism remains perhaps the least well understood (or most misunderstood) of these three, there are some insights emerging in the feminist work which are fundamental to current debates on the nature of the discipline.

This chapter outlines some ideas about the theoretical basis and empirical content of a feminist geography, suggesting some of the ways in which it is essential to ongoing discussion on the nature of human–environmental relations, specifically those being initiated by humanists and historical materialists. The first section proposes a general definition of a feminist geography as one concerned to understand the relationship between gender and environmental constitution. This is illustrated in a discussion of how women's activities have simultaneously altered gender relations and environmental relations, specifically focusing on the ways in which women earning money at home are altering the form and meaning of the home and community. Finally, this discussion is utilized to outline some implications of incorporating a feminist perspective into disciplinary discourse as a whole, arguing that this implies a further *rapprochement* between humanism and historical materialism.

FEMINIST GEOGRAPHY

Feminist geography, like feminism as a whole, is not 'only' about women. It is about the social and environmental processes whereby the two genders – women and men – which make up the category 'human' are constituted, reproduced and changed.[3] Feminist geography is about the way in which gender is constituted and how this relates to the constitution of the environment.

The development of a basis for feminist geography – concerned with the relations between gender and environment – had two initial components. First it was necessary to disaggregate the category 'human' and to reconstitute it as a category which included the activities of both genders. Second, it was necessary to disaggregate the category 'environment' and to reconstitute it as a category which was sensitive to alterations in gender. The following section outlines these two components as the theoretical underpinnings for empirical investigation in the subsequent section.

Feminist exploration in geography, as elsewhere, began by breaking down the ahistorical category 'woman' (and concomitantly 'man') and asking – what kinds of social relations structured and changed gender categories? This social conceptualization of gender depends on seeing gender – those socially constructed activities and attributes ascribed to women or men – as historically mutable. Gender categories are socially and politically constructed, socially and politically contested and socially and politically changed.[4] A good deal of feminist analysis is concerned to understand the nature and process of these changes.

For example, the gender category 'women' in North American

and Western European society has altered significantly within the last 50 years, largely through changes in what women are doing. A generation ago, a 'woman' was someone imbued with an overwhelming maternal instinct who confined her person to the home and her activities to nurturing an unrestricted number of children and a breadwinner husband. A 'woman' is now an energetic actor, capable of reconciling two demanding jobs – one at home and one in the wage labour force – and doing so rationally and cheerfully.

Some of this alteration was media stereotyping. But it was and is also a reflection of very real and substantial changes in the activity patterns and expectations of women and men. These altered activity patterns were themselves reflections of general social changes, changes which included relatively safe and reliable female-controlled contraception, access to better domestic working conditions, and a rise in female wage labour force participation. And although the women who organized for better fertility control, better housing and access to education and jobs did not, on the whole, consciously see the implications of their actions as leading to changes in the definition of 'woman', this was the necessary outcome of their new activities, as well as the precondition of further campaigns in these areas.

This constant change of gender categories proceeds in a reciprocal manner. Changes in the activities and attributes which make up 'woman' imply, result in and respond to changes in the activities and attributes of 'man'. The change from 'total mother' to 'superwoman' was accompanied by a concomitant change in 'man': from the sole breadwinner with his mind on the market and the sports page to a family companion and partner. The historical constitution and change of the two gender categories 'woman' and 'man' in combination, constitute and change 'human'.[5]

For feminists, therefore, to be human is not a static thing. It is a changeful and active process, altering through alterations in the activities and attributes which constitute 'woman' and 'man'. Just as the activities and attributes contained within the permeable boundaries of gender categories are not static, so the sum total of human activities and attributes is not static. Alteration of gender categories does not proceed through the transfer back and forth of a fixed set of ahistorical human characteristics from one gender to another. Change of the androgynous human is a creative process. Like all creative processes, it involves, at times, the destruction of (or failure to reproduce) some attributes in the creation of new ones. For example, men did not take on the roles of instinctively guided social maternalism as women failed to reproduce these, although fathering did become a more central and qualitatively different social institution. Similarly, the dual-role superwoman is, in many respects, a new historical creation. Gender, as the essential component of 'human', is permeable as well as mutable.[6]

A feminist geography suggests first that we look at 'human' as a category constantly changing as women and men alter gender attributes through altering their activities. It suggests we see that constitution of 'human' as an active and often conscious process. But the constitution of 'human' is only one half of human environmental relations which forms the basis of geographic enquiry. The second requirement of a feminist geography is that we develop a concept of environment which allows us to see the change of gender relations relative to environmental relations. This involves breaking down the category 'environment' and seeing environments as changing sets of resources appropriated in historically variable ways, these modes of appropriation altering through alterations in space and time patterns. Such a conceptualization is implied by the concept of 'human' outlined above.

A feminist geography therefore suggests that we see the environment, too, as an active, frequently political and sometimes conscious creation.[7] And it suggests that we see the constitution and change of an androgynous human and the constitution of environment proceeding simultaneously, in a process of interaction. In analytic terms, gender becomes a space-structuring force; environment becomes a component of gender constitution. The link between 'human' and 'environment' is the active use of time and space which simultaneously alters gender and alters the social environment.

Over the past century, people's space and time have become increasingly separated into 'work' for wages and 'life'. This has been a function of, and has reinforced, a growing separation between wage workplace and home, on a continuum which runs from the pre-industrial home-workshop to the suburban home distant from centralized industrial-commercial firms. This spatial separation has been correlated with an increasing differentiation between the gender categories woman and man. Yet this separation and the form of this differentiation are in the process of fundamental alteration and potential convergence in the current period of economic restructuring. The following section examines the interrelated development of this spatial separation and gender differentiation in 20th-century cities. It then explores how one group of women – those earning money at home – are actively breaking down the separation between 'work' and 'life' and, in thus changing their space and time patterns, altering the context and content of both 'human' and 'environment'.

CHANGING THE RELATIONS BETWEEN WORK AND LIFE

The 19th-century 'woman question' was resolved by confining most women's persons and defining all women's natures in terms of the home and community. The newly separated residential suburb had collected together all requisites of social reproduction and areally distanced them and the women who carried out domestic work from production.[8]

Throughout the 20th century, women lived and worked within the bounds of the gender-typed, separated city, invisibly organizing to improve their working conditions.[9] The bounds extended spatially into more extensive and ever more far-flung suburbs. The suburbs 'worked' for that brief historical period between the 1930s and 1950s when some men (it was assumed all men) had secure, remunerative jobs, and most women (it was assumed all women) could devote all their time to planning around and caring for the home and family. And their working assumed and reinforced this gender-typed separation. Male wages were sufficient to purchase a home and most of the raw materials to run it. Full-time female labour was required to actually shop for these materials and to turn them into the components of family life. Neither alone was sufficient. The social definition of 'woman' and 'man' became strongly imbued with connotations of natural differences which led to social reciprocity and consensus.

By the 1960s, however, the maintenance of this form of home came to rely increasingly on two incomes. Growing numbers of women took on dual roles, and the bundle of spatially reinforced gender differentiation began to unravel. The activities appropriate to women came to include earning money as well as providing goods and services in the home. Just as the suburban solution depended on and reinforced one set of gender roles – the full-time domestic worker-mother and the full-time breadwinner – so the maintenance of temporally and spatially separated homes and workplaces depends on another set, primarily women's dual roles (and to some extent, companionate marriages). But women's dual roles are themselves problematic.

Whereas the 19th-century 'urban question' had been primarily focused around inadequate conditions for reproduction in the central cities, the solution to these problems – the extensive,

distant and expensive suburb separated from wage workplaces –
becomes the mid-20th-century 'urban problem', a problem which
is especially acute for that growing proportion of the labour force
who are also domestic community workers. Work in the home,
the place associated with leisure, is not seen as real work, nor are
the home and neighbourhood designed as workplaces. What
women do in these places is invisible in the public sector, as it was
meant to be. Employers make few concessions to the fact that
women have other jobs. As it is assumed that someone is looking
after the children in the private home, childcare is not a public
priority. And at the end of a long day, having got her family off
to school, to childcare, to work, travelled to a wage job herself, to
shop, to see a child's doctor or teacher, the woman comes home
again to a domestic workplace which requires hours of hard work
and complex planning to keep running.

In response to the fact that women are working at the
intersection of home and work, providing not only resources in
kind in the home and community but also monetary resources,
they are creating services at the intersection of home and work,
services which bridge the two spheres and fill the formerly
invisible gaps in the separated city. As employment opportunities
and flexibility decrease with economic restructuring, and as
government restraint programmes cut back the already inadequate
social services available to families, especially to wage-earning
mothers, these small and socially unacknowledged services at the
intersection are becoming increasingly important to the economic
and social survival of households.

A growing number of women, or at least an increasingly
obvious group of women, begin to extend their home roles to
earning money at home.[10] Homeworkers are a heterogeneous
group. They are doing a wide variety of things; a recent study
identified over 200 types of home-based businesses (Johnson
1984). These range from international import-export businesses
and internationally renowned artists working from home studios
to the woman down the street who crochets dolls for sale at local
craft shows or minds her friend's children a few hours a day. But
whatever their differences, all these women have two things in
common. They have all adjusted their organization of time to
include work in the home for pay as well as providing goods and
services to their families. And they are all attempting to use the
resources available in the 'private space' of the home and
community to carry out a new set of activities.

The rigid, spatially reinforced separation between wage workplace and home – the concomitant basis of clearly demarcated

gender roles – was severely weakened by women's dual roles. It is being further dissolved by the adaptation strategy homeworkers are developing in response to economic restructuring and to the problems of maintaining dual roles. Homeworkers' adjustment of their space and time has implications which are potentially significant for restructuring the relation between home and work and between women and men.

Homeworkers: creating new urban resources

In an attempt to ascertain some of these implications, I interviewed independent homeworkers in the Trail-Nelson area of the interior of British Columbia and in the Kingston area of Eastern Ontario. I asked them not only what they did and how, but also about how their homework affected their relations to the resource base of the home and community, including their networks of family and friends. (Table 2.1 provides some information on the women to whom I spoke.)[11]

In general, I found that these homeworkers were utilizing existing local resources, primarily their homes and communities, and their given networks and skills. They were doing so in the context of specific environmental constraints. Many lacked viable employment alternatives, either because of their domestic responsibilities or lack of available jobs. And while most earned relatively

Table 2.1 Characteristics of homeworkers[a]

	Kingston	Trail-Nelson	Total
Total sample	59	63	122
Child caregivers	27	23	
Craftworkers[b]	32	40	
Number who had children at home	32	38	70
Average age (in years)	38	36	
Average family income (in thousands of dollars, Canadian)	28	24	

Notes:
[a]Interviews took place between January and June 1985.
[b]This group covered the range noted above, from artists with international reputations to people who produced 'in their spare time'.
Source: Questionnaire survey.

Table 2.2 Homeworkers' gross earnings as a percentage contribution to total family income

Earnings as % of total	Homeworkers		
	Total number	%	Cumulative %
Less than 10	7	5.7	100
11–25	36	29.5	94
26–50	40	32.9	65
51–75	13	10.7	32
76–99	12	9.8	21
100	14	11.5	12
Total sample	122	100.1	

Source: Questionnaire survey.

little from their homework, the job often provided a substantial part of family income (see Table 2.2).

Within these constraints, however, the social relations involved in attempts to utilize a limited set of resources designed for one purpose – private family life and individual consumption – in order to carry out a related but different purpose – the provision of a public service and the gaining of a livelihood – are causing homeworkers to alter the environments in which they work. They are using their given, limited environments and skills in new ways, principally by converting or redesignating resources formerly seen as private into public or collective productive resources. In the process, they are extending these resources, and often creating new resources.

This process of extending and creating resources at the interface of home and work has three implications for women's activities and for domestic community life. First, these women are developing new sources of employment, providing new social areas for 'work' and thus contributing to redefining the nature of work. Second, they are providing services to other members of the community, thus contributing to restructuring domestic-community working conditions, and to altering the nature of reproductive activities which go on there. Third, they are altering the balance of resources in kind and monetary resources which households need to survive, and thus changing the relations between home and wage workplace. In the following, each of these is examined in turn.

Table 2.3 Homeworkers' training

	Child caregivers	Craftworkers	Total
Total sample	50	72	122
None[a]	12	9	21
Seminars, drop-in courses	21	16	37
Courses in their area of homework	30	61	91
Other, related courses	17	14	31
Professional qualifications	21	20	41

Note:
[a] With the exception of 'none', categories are not mutually exclusive.
Source: Questionnaire survey.

First, women – caring for each other's children, sewing each other's drapes – are creating their own jobs. In the process, they are redefining and extending the sets of skills which are considered worthy of payment. These women are taking their 'natural' parenting skills, their homemaking skills or their experiences in home manufacture and crafts and redefining them as marketable skills. The private experience they have gained as mothers and housewives has been converted into a set of defined and valued skills with which to make a living. And the majority are attempting to extend their personal skills. Most have had some 'outside' training to enhance these skills immediately prior or subsequent to starting homework. This training varies from professional qualifications to informal drop-in or. self-organized collective work sessions (see Table 2.3).[12]

Most of these homeworkers have also redesignated networks among friends and neighbours as 'working' networks. This takes a variety of forms, based around formal or self-created organizations, or around informal *ad hoc* meetings of homeworkers with similar interests. Such networks were not only sources of contact, advice and assistance, but also often acted as referral and in some cases monitoring agencies. These networks acted to self-regulate and to disseminate information about the quality of goods or services provided by the homeworkers.[13] In developing and utilizing these networks, homeworkers are providing themselves with some of the protection and services of a combined trade union and professional association. They are also attempting to increase the social recognition of their skills through involvement in professional associations ranging from nationally organized groups to *ad hoc* neighbourhood networks.

Homeworkers have also altered and extended pre-existing networks with family. Most indicated that family members often helped with their work, either through directly working on the money-earning activities or in taking over other domestic tasks.

Second, homeworkers are providing direct 'public' services to other members of the community, services which, as noted above, are becoming increasingly central to domestic community life, and are often the only alternative available. And this has the effect of altering the conditions, and the environment in which domestic community work goes on. For homeworkers, both the household and the neighbourhood become workplace as well as living space, simultaneously public and private space. In fact, the division into public and private becomes an increasingly meaningless one. For homeworkers, the formerly 'separate' activities involved in maintaining the household become increasingly incorporated into the activities involved in earning money, and their content merges. A childminder cooking lunch for her own children as well as those she is paid to care for is simultaneously doing a 'labour of love' and wage-earning labour. In many cases, the places where these two sets of activities go on are the same, or parts of the home are set aside primarily or exclusively for money-earning work. Women working at home are in fact redesignating, and in some cases redesigning, their homes as places to earn a living as well as to 'live'.[14]

Similarly, homeworkers are redesignating their neighbourhoods as workplaces, assessing the local and wider communities in terms of demand for their product or service and in terms of facilities to assist their work.[15] Women using their homes and neighbourhoods as workplaces are also actively redesigning these environments. Not only do the houses of many homeworkers include spaces set aside for the activities of homework, but these women are creating, in concrete terms, new public spaces; milieux where both 'private' domestic-community work and homework go on collectively. Such spaces include the sites of playgroups and drop-in centres, attended by both mothers and homeworkers, the crafts guilds which include both 'hobbyists' and homeworkers, co-operative craft shows, stores and networks which purchase materials and equipment for both homeworkers and 'private' domestic workers.

Therefore, these networks and spaces not only provide services to homeworkers, they are also restructuring the conditions under which women with dual roles or no paid jobs are carrying out their 'private' domestic community work. A regular part of parenting increasingly includes attendance and some organizational work at a co-operative playgroup. A regular part of maintaining the home may include attending informal classes on craft manufacture and the purchase of materials from co-operative networks.

Third, women earning money at home are restructuring the balance between domestic work – producing free services and goods in kind – and wage work – producing money for the purchase of services and goods. This alters both the value of labour power and the conditions of its reproduction.

Wages have never been sufficient for most people's subsistence. Subsistence has always required some form of domestic work, and for most, some form of family and community-based mutual aid. Women in the community have a long tradition of organizing community mutual aid, based on attempts to compensate for the quantitative and qualitative differences between the male wage and the needs of the family.[16] But such organization has necessarily become more dense and more formal as an increasing number of women, especially mothers, entered the wage labour force. A loose network of extended family and friends could meet the needs of a community of full-time housewives-mothers on an *ad hoc* basis. However, more regular assistance, more often co-ordinated by exchange value relations rather than use value relations, became necessary to meet the needs of a community of dual role women. The forms of mutual aid provided in the community thus become more central to the exchange process between labour power and capital as a growing percentage of the labour force define themselves as domestic workers as well as wage workers.

Wage work patterns are also influenced by the homework sector. A socio-economy where the maintenance of family life requires both adults to be economically active tends to break down the barriers, real and analytic, between wage work and consumption on a number of levels. And this breaking down is extended as the services necessary to maintain the labour force include more and more aspects of the 'social wage', and issues such as childcare become workplace issues.[17] Formerly 'private' activities like childcare become workplace issues not only in a formal sense of being union bargaining matters or incentives to attract labour, but also in the sense that the availability, affordability and accessibility of childcare help determine the availability of parents for wage work. The location and accessibility of childcare influences the range of places parents (especially mothers) are able and willing to work. And as the majority of childcare is provided by home caregivers, the size, location, cost and quality of this sector becomes intimately bound up with both the availability and the satisfaction of the wage labour force, especially the participation of mothers in the labour force.[18]

The homework sector, as a community resource, also influences the amount of money a family needs in order to reproduce itself; that is, it influences the balance between use values and exchange values in the household budget. The homework sector often provides alternatives to formal market services, not only in the case of services like childcare, typing or household maintenance, but also in the form of alternative sources of household goods. Not only do homeworkers manufacture goods for sale, the networks of all homeworkers are supplemented by lively barter relations. This includes both barter in kind (e.g. pottery for clothing), and also services for goods (e.g. babysitting for paintings, craft lessons for garden produce). In addition, the co-operative networks set up to purchase materials for craft manufacture, utilized by both hobbyists and professionals, also provide alternative sources of raw materials for household manufacture.

Homeworkers are thus involved in altering the nature and content of exchange relations, sometimes on a small scale, as with most craft manufacture, sometimes, as in the case of the home childcare sector, dominating the field. In some respects this takes the form of an alternative economy. In most respects, however, it is an alternative intimately bound up with and changing the nature of the mainstream socio-economy.

I stated at the outset that a feminist geography is based on the understanding that gender and environment alter simultaneously, in a process of interaction, and that the link between an androgynous human and the environment is the active use of space and time by women and men. Women's dual roles can thus take a multitude of forms. Women can work in two separate and functionally disparate places, attempting to bridge constant discontinuity in their daily lives, largely through individual effort and the use of a range of use and exchange values available in their communities. Alternatively, they can obtain a variety of resources in a variety of ways within the home and community, and thus alter not their location, but the social meaning and function of home and work. However they carry them out, women's dual roles are not a radical break with the past, but another adjustment of their use of space and time in order to maximize access to resources.

In their daily activities, women make multitudes of small, daily decisions and choices, attempting to adjust the time and space for care of children, for domestic community work, for money-earning work. All of these arrangements must take into account the fluid constraints imposed by the specific balance of resources and family needs. Women's allocation of space and time must attempt to balance the resources the family needs – the amount of money, the number of commodities and services, the amount of time from the mother, the wife, the cook, the wage earner – which is required to achieve a desired family lifestyle. All of these adjustments have had specific effects on the nature of capitalist society, and more immediately but less obviously, on the nature of both the home and community and of gender relations within this society.

Homeworkers, in creating a new social space at the intersection of home and wage workplace, are thus part of a radical, rooted continuity. This is a continuity which stretches both across time into the social history of pre-industrial society and across space into the economies of contemporary third world countries (Oppong 1983). But saying it is not new does not imply that its implications cannot be new, nor that the social milieu it creates cannot be the source of new and potentially consciously creative social and gender relations. The implications can be both exciting and horrific (Gershuny 1978, Mattera 1985, Townson 1984); they are certainly worthy of serious analysis.

Carrying out such analysis requires not only that we recognize the reciprocal alteration of gender and environment, but also that we define the 'necessary relations' or the essential, definitive parameters of gender constitution and change, which in the case of feminist geography are simultaneously the parameters of environmental constitution and change. If we see human action as the link between gender and environment, we must address the question of the essential relations toward which such action is directed. The final section of this paper returns to feminist analysis as a whole in the discussion of some of the implications of this new focus for the study of human–environmental relations.

RESTRUCTURING SPACE AND GENDER

Both socialist and radical feminists have defined, albeit in very different ways, the essential parameters of gender constitution as the relations between the production of society's goods and services and the reproduction of people as biological and social beings.[19] And while radical feminists concentrated on the politics of reproduction, socialist feminists attempted to articulate this understanding to broader theories of social change, primarily, although not exclusively, historical materialism.

In so doing, socialist feminists have not just adopted, but also adapted, historical materialism. Including gender is not just a matter of an empirical addition at the historical level. It involves, or more accurately makes evident the need for, a shift in our analytic focal point. Women in capitalist society (and by implication men) are defined from, live and work in, and analyse themselves from an inherently contradictory material position. Women are defined as responsible for essential social work – mothering and caring for adults – what is commonly called reproductive work, in a society where power and analytic categories derive primarily from the productive sphere. Like class relations, but in a more obvious way, gender relations are constituted not only by relations of production, but also by the dialectically related relations involved in the reproduction of people. Seeing the articulation of gender relations with environmental relations means focusing on the activities involved in the production of goods and services, with reference to or simultaneously with those involved in the biological and social reproduction of people.

As a consequence, socialist feminist analysis focuses not primarily on production, not primarily on reproduction, but on the relations between the two. The nature of this intersection is seen to define gender relations. These are the essential social parameters which analytically encompass women's activities, those activities which constitute and change gender and environment. The relations between production and reproduction are thus the 'necessary' parameters in defining gender and its interaction with environment.[20]

But the proposed articulation of gender and environment implies even more than a change in the initial focal point from which our categories are formulated. It implies a new political focus, one concentrating on understanding the content of categories and the human activities which make up this content.

We cannot focus long on the relation between production and reproduction without moving a long way from both the bad old days of economic determinism and from the seductive Althusserian moment – the theoretically comfortable but politically irresponsible view of people as bearers of relations. In part this is because such a focus breaks down the neat conceptual divisions between production and reproduction; between work and non-work or life;

between productive and non-productive, even between class and gender, or structure and agency. It forces us instead to see people's activities anew, conceptually unclad and undisguised, so to speak. In part, looking at reproduction as analytically important, as politically pertinent, makes a focus on content unavoidable. There is little else to see apart from the small, 'insignificant', 'common-sense' details of keeping people alive and socially integrated in a capitalist society. And as these become evident, they mingle with or even crowd out the abstract questions of tendencies and structures which formed the basis of positivist geography and which still linger on in some contemporary work.

. So a focus on the intersection of productive and reproductive activities tends to create an analysis which is full of content, which is concerned with the kinds of interpersonal relations through which we get by day by day, and which is concerned with how we can change these. It suggests that we re-centre our attention on the taken-for-granted details, the small human actions which constitute and change our categories, our tendencies, our structures, that we look for social changes at the level where they are created, in human lives, as well as at the higher levels of abstraction with which many of us have become so expert at conjuring. In short, this focus is one of the major analytic bases for the emphasis, in feminist political actions, on prefigurative politics, on building a new society from the bottom up, so to speak (Rowbotham 1972, Rowbotham *et al.* 1979).

In general terms then, socialist feminism remains profoundly committed to historical materialist analysis – seeing life in capitalist society as structured by our relations to the means and processes whereby we produce goods and services and reproduce ourselves as biological and social beings.[21] It is also profoundly humanist in being centred on human agency, concerned with daily actions as the basis of social change.

The preceding discussion therefore suggests that the development of feminist geography has implications both for the theoretical-methodological basis of the discipline, and for the nature of our empirical work. Feminist geography was initially defined as a study concerned to understand how changes in environmental relations – changes in the way people use and create space and time – interact with, help structure, reflect and reinforce the social definition of the gender categories 'woman' and 'man', and the relations between women and men. It is concerned to understand the ways in which the activities involved in being a man or a woman in society relate to appropriation of and activity in time and space. It implies an interactive and creative relation between gender and environment, and much current research is devoted to understanding the nature of this relation.

It was suggested that the extension of this perspective to the discipline as a whole and its contribution to current debates on human environmental relations directed our attention to two areas. First it implied a disaggregation and reintegration of the 'human' half of the human–environmental equation to include the diverse and often oppositional creativity of the two genders. Second it implied a focus on environment as actively constituted and changed simultaneously with gender, through alterations in the activities of women and men. This latter was seen to involve a focus on the interrelated activities of production and reproduction. Empirical discussion of the historical nature of this relation revealed not only its variability but an emerging tendency toward breaking down the spatial and functional separation between productive and reproductive activities. Although the activities of homeworkers in creating this breakdown were in one sense reproducing women's domestic community skills, they were simultaneously altering the basis of gender constitution.

A geography which asks us to study the reciprocal relations between the constitution of an androgynous human and the constitution of environment asks a great deal, nothing less than the disaggregation and reintegration of the most fundamental concepts in the discipline. This, however, is no less than the changes which inspired feminism require. Within the last generation, women have seen tremendous alterations in their daily lives and in the conditions under which resources for these lives are provided. They have also seen the image of woman change radically, and seen and experienced feminism redefine 'woman' from a state of being to a political force. But the potential implications of the restructuring of productive and reproductive activity which homeworkers are creating are perhaps even more profound, both for our environments and for gender relations.

NOTES

A different version of this paper is reprinted in Hamilton & Barrett (1986). I would like to thank Roberta Hamilton, Audrey Kobayashi and Alan Nash for comments.

1 Some studies of women's complex travel and activity patterns include Fava 1980, Michelson 1985, Palm & Pred 1978, and Wekerle 1984.
2 For a review of the geographic literature, emphasizing the empirical

work, see Zelinsky *et al.* 1982. Interdisciplinary collections on women and environment include Keller 1981, Stimpson *et al.* 1981, Wekerle *et al.* 1980. Some discussion of the relations between feminist analysis and geography as a whole are Bowlby *et al.* 1981, McDowell 1984, 1985, Mackenzie 1984, Mackenzie & Rose 1982, Monk & Hanson 1982, Women and Geography Study Group of the Institute of British Geographers 1984.

3 Not all feminists would agree with this statement, nor with the subsequent analysis of 'human' which flows from it. Certainly, radical feminists, who have been more concerned with human nature and with biological and psychological questions than have socialist feminists, see the project of feminism as more concerned with women's liberation, and the nature of human, especially of 'man', as more immutable. See, for example, Miles 1985.

4 This understanding of gender depends on the distinction between sex and gender. Sex categories are those defined by biological differences. Gender categories are social constructions, historical and cultural definitions of the social and physical capabilities and the appropriate activities of 'women' and 'men', and of the nature of femininity and masculinity at any point in history.

   The dividing line between sex and gender is not a rigid one. Gender categories are obviously influenced by sex categories, but are not fully defined by them. Many aspects of biological sexuality are socially influenced and gender categories can be occupied by people of either sex.

5 The change of gender categories is always a social and often a consciously political process, one embedded as deeply and ubiquitously in the social fabric as is the division of human into 'woman' and 'man'. Because it is so deeply embedded, such changes are often invisible, or, if changes in gender categories are recognized, they are assumed to be confined to women. Although changes in 'woman' have perhaps been more obvious, largely because feminists have concentrated on exploring and analysing women's lives and because of the frequent confusion of gender category 'man' with 'human', it is obvious that the gender category 'woman' does not change in isolation.

6 Neither does change in 'woman' proceed through opposition to change in 'man', in the sense of one gender gaining new options by taking them away from the other. This common idea is based on an odd view of humans as historically endowed with a given set of social 'goods' which means gains can only be achieved through reciprocal losses by another. Gender change often is an antagonistic process, but it is not inherently so.

7 This conceptualization of people and environment is related to developments in historical materialist and humanist concepts of human–environmental relations which have developed within or have influenced geography. There are parallels with Bertell Ollman's work on the creation of nature and human nature through environmental

appropriation in work (Ollman 1976), and with Anthony Giddens' ebullient discussion of space and time (Giddens 1979, 1981). In geographical work, this concept is connected to work by Ted Relph (1981) and by David Ley and Marwyn Samuels (1978) in their concern for intentionality and created meaning in environmental relations, a theme which is also central, although from a different perspective, in Andrew Sayer's (1979) discussion of people and nature.

8 This relatively new historical creation of the 'private' home and community and the full-time mother–housewife was reinforced and perhaps inadvertently naturalized in urban science. Because the home and community were concretely, undeniably *there*, separate from productive activity and market relations, and associated with emotion, leisure and sexuality, urban analysts built a science around this division, and tied it together with a 'science' of the natural differences between women and men. The separated city was apparently implicit in human nature. The separation of home and community from 'work' permeates the whole history of human geography, perhaps most especially urban geography. It is evident in historical materialist work which, in this sense, is still manfully maintaining the tradition of separating social and economic concerns, a separation stemming from the division between the social Darwinism of the Chicago school urban ecologists and the neoclassical land economists. It is only recently, primarily through the explorations of feminist historians and environmental scientists, or through work inspired by these, that the mutable nature of the home and its occupants has become evident.

9 Such organization includes the establishment of Women's Institutes and Co-operative Guilds to improve education and consumer services, groups which pressured for better housing, for family allowances, for nursery schools and extended welfare services. See Mackenzie & Rose 1983.

10 It is difficult to establish with any precision the number of people working at home for pay. According to estimates, between 1980 and 1984 the self-employed category in Canada, which would include the homeworkers interviewed in this study, increased at a faster rate than did the numbers in paid work as a whole. Rates of growth were especially high for women. For example, between 1980 and 1982, numbers of paid workers of both sexes decreased by 4.7%, numbers of self-employed women increased by 11.3%. Between 1982 and 1984, numbers of paid workers increased by 5.9%, self-employment by 9.5% and self-employment among women by 20.5% (Levesque 1985). In the United States, the Small Business Administration estimated that in 1984 more than 2,000,000 American women were running businesses from their homes (Johnson 1984).

11 All the homeworkers interviewed were 'self-employed': running independent businesses rather than working as home-based contract workers.

12 For child caregivers, training ranged from seminars and drop-in centres co-ordinated by Family Day Care agencies with the assistance

of statutory bodies or by the Board of Education and Recreation Commissions, through formal adult extension courses in first aid, recreational training and creative education to certificates in early childhood education. For craftworkers, training ranged from seminars given by local artists and crafts councils to university or art school extension courses or degrees in fine arts. Many craftworkers had also travelled and attended numerous shows and conferences.

13 Most frequently mentioned networks for caregivers included drop-in centres for home caregivers, local playgroups, regional or national childcare groups. For craftworkers, most frequently mentioned groups included local craft councils or arts councils, Women's Institutes and *ad hoc* local groups.

14 The majority of homeworkers had carried out some renovation, the amount varying positively with the time they had been doing homework and with their satisfaction with the job. Such renovation varied from redesignating and adding small equipment or minor furniture to parts of rooms to the addition or finishing of rooms or major yard work.

15 For example, home caregivers discussed their neighbourhoods in terms of their capacity to provide safe, accessible play space and public transport or settings for outings. They also assessed the neighbourhood in terms of the number of local families requiring childcare and in terms of the ages of children, thus measuring the actual and potential market demand for their services. Craftworkers assessed communities in terms of potential markets as well as in terms of available local outlets for their products.

16 These range from *ad hoc* sharing and borrowing networks through formal women's groups to taking in boarders (see note 12. Also Bradbury 1982, Fox 1980, Tilly & Scott 1978).

17 Not only have they become, to some extent, union issues (see Coote & Kellner 1980, Fitzgerald *et al.* 1982), but in some cases childcare is located in the workplace (Friendly & Johnson 1981).

18 In 1983, of children receiving some form of non-family care, home caregivers in Canada were providing care for 84% of children aged two to six and 95% of children under two (Canada, National Health and Welfare 1984). On the USA, see Fosberg 1981.

19 Socialist and radical feminists are, in general terms, the two major feminist frameworks which have emerged in answer to the question of the essential relations toward which political strategy should be directed. Radical feminism sees the conflict between genders as the primary historical and social conflict, one which has biological and psychological roots reinforced by social practice. Socialist feminism focuses on the institutions and social practices of capitalism, or other modes of production, as these constitute and change gender. Socialist feminists see gender and class as constructed through historically concurrent although not identical processes. For further discussion of these two frameworks, see Barrett 1980, Eisenstein 1979.

20 These are necessary relations in any society, in that one presupposes

the existence of the other and their form and content are defined by their mutual relation. Renate Bridenthal (1976) calls this a dialectic. It is not necessary that their relation take any particular form, nor that they contain any given content, nor that their activities be carried out by a particular gender.

21 Considering relations to the means and processes of reproduction makes it evident that many of the results of wage earners' production form the *means* of reproduction. For domestic workers (*as* domestic workers) the relations to production are less important than the wages available and the goods available. Her wage levels, and her husband's, structure her conditions of reproductive work, determine her access to means of reproduction, and influence her process of reproduction. Of course, the relation is reciprocal. This may be at least a beginning at unravelling the relation of 'gender and class constitution' – a question made artifically thorny by discussion which utilizes pre-feminist categories and appears to operate at every level of abstraction except that of historical tendency.

3

Theory, hypothesis, explanation and action: the example of urban planning

JEANNE M. WOLFE

INTRODUCTION

Urban planning, like many of the helping professions, is a consumer of knowledge, ideas and theory from the academic disciplines. Geography, being concerned with spatial distributions and human–land relationships, has been and is a major source. In fact many professional planners first graduated in geography. The contemporary evolution of geographic work is thus of direct interest to planners, not only as a source of ideas but also as a critique of their work.

The same observations could be made for other interventionist occupations where geography and geographers are important. These include conservation and environmental work, resource planning and management, regional analyses and prediction, transportation routing and mapping programmes of all sorts. The search for relevance in geographical research and emergent theory must therefore include a look at how it is used in other fields.

The purpose of this chapter is to review the development and role of theory in urban planning, and to show to what extent it borrows from geography and to what extent it diverges. It is divided into four sections. The first examines the general problem of theory and how it can be judged useful in planning. The second proposes a working classification of planning theory in an attempt to bring some order into a rather jumbled field, and to make subsequent discussion clearer. The third section then reviews the positivist approach along with the humanistic–materialist debate in relationship to this classification and demonstrates the utility and limitations of the rational model. It endeavours to show that planning theory is the product of the broader intellectual, social and economic climate of the times, and that theories linger on, coexisting rather than replacing each other. A fourth and concluding section reflects upon the divergent nature of professional and academic work and speculates on future directions.

THEORY AND ITS VALIDITY

Theory is an abstraction of reality that gives the student of urban affairs both a shorthand and a framework within which to work. Theory can be implicit, as it is in most of our daily lives, or explicit, as encouraged by academic convention. It can be simple and it can be complicated, although as an abstraction, one would expect a tendency to the former. It can operate on many levels or within closely circumscribed parameters. It is always based on assumptions, which in themselves should form part of the theory, but are often not stated. Theory is often made unnecessarily difficult by the use of obscure language which hinders debate between people of different education, culture and background.

In its broadest sense the word theory can be used for any collection of ideas, concepts or intellectual processes. In its supposedly scientific sense it has three definitions: first, a plausible acceptable explanation of observed phenomena, the Baconian inductive method of logical inference; second, an hypothesis assumed for the sake of argument or investigation to be subject to empirical testing and, in the Popperian mode, possible falsification; and third, a guide to action. The first two, while apparently contradictory, are usually used together in a mixture of inductive generalization and deductive reasoning in most social science research. The use of the third distinguishes urban planning from urban studies, human geography, sociology and other academic disciplines. The urban planner has to look forward, has to propose interventions in the human environment, and has to move beyond prediction into normative theory and action.

This fundamental difference between the academic discipline of geography and the profession of planning often seems to escape the casual observer. The urban planner is oriented toward active intervention in the future through planning, design and the shaping of public policy. The responsibility of the geographer on the other hand has been to the discipline of geography and the advancement of knowledge. Thus while geographers might study the effects of urban planning on the landscape (see Hall 1982) the reverse is not true. Geographers' work is judged by their peers whereas planners' work is judged by the general public and/or their clients.

This means that, in general, research in urban planning, unlike that in the social sciences, is geared to specific problem solving rather than to the general advancement of knowledge. The reason for doing planning research is to suggest a solution, and any scanning of thesis titles or planning journals will show that research initiatives are mostly opportunistic.

When the objective of theory formulation is not simply explanatory, but includes its use in planning for the future, we are concerned not only with the conventions of theory formulation, but also with specifying performance criteria. Somehow theory must meet the requirements foreseen to be important in the field. This immediately presupposes some system of values, if not divine guidance. There are at least four criteria of adequacy which have been proposed for judging if a theory is useful coming from the positivist school. These are: (1) a theory must have a dynamic aspect; (2) it must be susceptible of empirical verification; (3) it must have internal logic and consistency; and (4) it must not be so abstract as to have no relation to reality (Ackoff 1962, Chapin 1965). To these, the humanists would add social usefulness and the Marxists, historical continuity, internal logic through dialectical reasoning and a broad societal context in the political economy sense. The Frankfurt school represented by Habermas would include communicative competence characterized by comprehensibility, truthfulness, sincerity, and legitimacy (see, for example, Albrecht & Lim 1986, Harloe 1977, Ley 1978).

THE CLASSIFICATION OF PLANNING THEORY

At the outset it is useful to divide planning theory into three groups: (1) substantive theory, relating to design and policy in urban planning; (2) theories of explanation of urban phenomena; and (3) theories of process, that is, theory of procedure and implementation. It might be noted that most observers have simply made a distinction between theories of the planning process and theories with which planning is concerned (Friedmann & Hudson 1974, p.2, Hightower 1970, p.326). Faludi (1973a, b) calls these theory of planning and theory in planning, but this is not particularly helpful in sorting out the descriptive and explanatory from the normative and action guiding.

Substantive theory is that which has to do with creative ideas in policy and design in urban planning. It includes physical concepts such as green belts, new towns and neighbourhood units, legal concepts to do with property and regulation, aesthetic concepts such as fit and congruity, and social concepts to do with the ever-elusive public good. Substantive theory is concerned with normative ideas about the spatial ordering of the built environment. It is the oldest branch of planning theory, dating from the earliest days of civilization.

Substantive theory goes in scale from overall urban form (star, linear, rectangular, etc.) down to site planning and the form of individual buildings. It has evolved through military, political and religious considerations, and through economic, social and environmental analysis (Gallion & Eisner 1950, Hiorns 1956, Morris 1979). It has always contained a strong utopian ideal, and an unshakable belief that the urban environment can be improved (Rosenau 1959). Its major exponents have been, and continue to be, engineers, architects and landscape designers. Kevin Lynch (1981) has been the major contemporary contributor to this field, which is succinctly summed up in the title of his last book, *A theory of good city form.*

It has been popular, and sometimes with good reason, to criticize physical planning for its lack of human consideration (Goodman 1971, Jacobs 1961). However since the days of Ebenezer Howard (1898) and Patrick Geddes (1915), and as popularized by the English Town and Country Planning Association and the Regional Plan Association of New York, particularly through the writings of Lewis Mumford, social and economic theory have been major inputs into the design theory of the times. For instance, Howard's (1898) theory of garden cities seems to be largely remembered for its catchy magnets and diagrammatic lucidity. However, a casual reading of the text shows that the problems addressed included the overcrowded, insalubrious inner city, economic self-sufficiency, the recapture of increased land value by the community as a whole, neighbourhood units and community control. The last four are as valid today as they were then, and the focus of much contemporary co-operative theory and praxis.

Planning ideas and concepts are much affected by societal and technological change. The substantive theory in vogue at any given time is often the result of a mixture of invention, context and reaction to another set of theory. Opposing theories can be in fashion simultaneously. Thus Le Corbusier, thrilled by the development of the skyscraper, itself a product of elevator technology (high buildings had been known before, usually raising their heads to God), proposed the rebuilding of Paris as *La Ville radieuse* (1933). As a corollary, he denounced the burgeoning satellite garden city movement as an 'assassin' of the central city. At about the same time, Frank Lloyd Wright, conscious of the great open spaces of America, and the power of the automobile in the freedom it brings, was espousing 'Broadacre City' (1934).

The modification of substantive theory usually comes about when it is seen not to work in a particular context. Thus large-scale residential demolition for urban renewal (cut the cancer out), was shown not to solve problems for inner city residents, and has, at least in the Western world, been replaced by softer programmes of renovation and rehabilitation. The spate of residential demolition in Montreal, for instance, started in 1958 with the clearing for the Jeanne Mance scheme and finished in the post-Expo depression after 1967.

The second group, explanatory theory, describes and explains urban distributions and structure. It is the meat of most urban geography courses, although strictly speaking it is not all attributable to geographers. It includes such constructs as organismic analogies, concentric zone theories, sector and multiple nuclei theory, central place theory, economic models of spatial structure, density-distance theory of all sorts, wave theory of metropolitan growth, industrial location theory, CBD and commercial hierarchy theory, urban land markets, land use distribution theory, social area analysis, network and urban systems theory. Its heroes are household names that include Schnore, Alexander, Park, Burgess, Hoyt, Harris, Ullman, Christaller, Von Thunen, Weber, Losch, Isard, Berry, Webber, Clark, Pred, Alonso, Muth, Smailes, Dickenson, Chapin, Vance, Murphy, Shevky and Bell, Haggett, Harvey, Bourne (see, for example, Bourne 1971, Carter 1972).

Explanatory theory is frequently quantifiable, usually has dynamic aspects and is often used as a basis for building predictive models. It does not say whether any given distribution is desirable or deplorable. It is supposedly built up on value-free empirical evidence, although as Ley (1978) has shown, this may not be entirely the case. (C.P. Snow, as early as 1962, was arguing that all science is un-neutral.) Almost all of this group of theory comes from the social sciences, geography, economics, sociology and political science, and is the product of this century. Most of it comes from the positivist school of thought, although more recently some comes from humanist or neo-Marxian interpretations.

Explanatory theory has two major uses for the planner. First, as already mentioned, it usually forms the basis for predictive model building. Its use in demography, land use, regional economics, transportation network simulation and for commercial facility siting is well known. Second, it provides a shorthand for the description and imaginability of the city, a set of rough descriptive tools which aid in interprofessional communication.

The third major set of theory used by urban planners is procedural theory, that relating to the process of planning. In fact, in much contemporary American work, the term planning theory is used only to describe this group. A notable exception is the compendium of Melville Branch (1975), and of course the work of Kevin Lynch (1981), already mentioned.

Procedural theory deals with the nature of the urban environmental decision process – how it is and how it should be conducted. It borrows heavily from the policy, administrative and management sciences. Of the three groups it is the latest to receive scholarly attention. This essentially dates from post-Second World War times, although debate on the subject started as the question 'to plan or not to plan?' in the early days of socialism.

In this context it is worth noting that modern planning has oddly contradictory beginnings in socialist and capitalist nations, although they have subsequently come closer together. In post-revolutionary Russia, national planning was the key to restructuring; in fact it was to be the cornerstone of the new state. As such it came to be viewed askance by the West as part of the evils of communism, although after 1945 some countries, such as France, adopted it as a technique to further reconstruction (Cohen 1977). Similarly, newly independent colonies have used it widely to guide their development. Meanwhile, planning in England, Canada and America grew up as a pragmatic response to functional imperatives. This was first through public health concerns as developed because of the appallingly squalid conditions of the mid-19th-century industrial city, and later on as a reaction to the extraordinary conditions of the great depression and then the Second World War. Whereas early Soviet planning was largely to do with the location of industry and its infrastructure and the setting of production targets for the various agricultural and industrial sectors, in the West it was focused on housing, parks, schools, transportation systems and social services, that is the domestic rather than the industrial environment.

The great debate about the merits of adopting planning really marks the beginning of the articulation of procedural theory. Four writers typify this debate. For Von Hayek (1944), planning was *The road to serfdom*, and for Karl Popper (1944) it was the enemy of *The open society*, whereas Barbara Wootton (1945) maintained that there would be *Freedom under planning* and Karl Mannheim (1946) that there must be *Planning for freedom*. From this time on, attention has focused on what should be planned, how it is planned and how it should be planned. Scholarly work has looked at the methods by which plans are formulated, how decisions are made, how conflicts are resolved, how policies, plans and programmes are implemented and how they are or should be evaluated (Blau *et al.* 1983, Burchell & Sternlieb 1978, Clavel *et al.* 1980, Faludi 1973a, b, Paris 1982, Robinson 1972). With a few exceptions, geography has not been much concerned with this field.

This classification of planning theory has been sketched out in an attempt to bring order to a rather jumbled field. However, it should be noted that the groups proposed are not necessarily mutually exclusive. Theory relating to citizen participation, for instance, cuts across both the substantive and the procedural, whereas that relating to industrial location has both explanatory and substantive elements. The following discussion will focus mostly on the development of theories of process.

THE RATIONALIST APPROACH AND THE HUMANIST–MATERIALIST DEBATE

The rational model has been and remains the basic methodological approach of urban planning since its foundation as a profession at the beginning of the century. At the same time, because of its reform roots, like social work, it has always been ideologically bound. At the beginning it was strongly humanistic, then with the development of formalized positivist approaches in the social sciences it became more 'scientific'. By the mid-1960s the 'efficiency versus people' debate started to shift again toward people, and by the mid-1970s this was being reinforced by neo-Marxist analyses. The coexistence of several apparently contradictory streams of thought seems to be troubling to purist academics, but to the practitioner, accustomed to trying to reconcile competing claims, this does not seem to be as monumental a stumbling block. This section will elaborate these assertions.

Modern town planning emerged around the turn of the century as a cause, and later as a profession, as an outgrowth of a series of reform movements (Rutherford 1974, Stelter & Artibise 1977). First were the public health reformers who saw the problems of social inequality, the horrors of slum conditions in industrial cities, overcrowding, disease, contaminated water supply, inadequate sewer systems, all exacerbated by corrupt municipal administrations. Second was the 'city beautiful' movement triggered by the Paris exposition of 1886 and the Chicago World Fair of 1893, which showed that an urban environment could be glorious, uplifting, green and clean. Walt Disney was a small boy in Chicago who cried when the fair closed. Third was the conservation movement, precipitated by the fast disappearance of natural resources under reckless exploitation, a movement that grew to encompass people concerns and the parks and playgrounds movement.

Spragge (1975) has demonstrated how these social forces developed and how planning came to be seen as a panacea for all urban environmental ills. He demonstrates how these ideological origins can be classified into the earliest goals of Canadian planners, namely, efficiency, redistribution and an aesthetic ideal. From these coalitions of public health activists, municipal reformers, architects, surveyors and engineers, the profession gradually emerged (Gunton 1981). The Town Planning Institute of Canada was founded in 1919.

From these humanist beginnings, ideas of efficiency gradually took over. The idea of making the city more functional was essentially a prolongation of the urban reform movement. Efficient waterworks, utilities and transportation systems, avoidance of duplication, honest value for the expenditure of public funds, unwasteful land subdivision schemes and salubrious worker housing all became rallying points. Good comprehensive town planning, including orderly street layouts, pleasant parks, minimum standards for housing and the regulation of land uses through zoning would improve residential areas, increase land values and tax returns, and keep the population healthy and contented. That things did not work out that way is well known, but it is from this time forth that faith in the regulatory approach can be dated.

The earliest exhortation as to method in planning probably came from Patrick Geddes (1915) who summed it up as 'survey – analysis – plan'. This was certainly the method of the first professional planner brought into Canada from England, Thomas Adams of Commission of Conservation fame, and one of the founders of the professional institute (Armstrong, 1968). The idea of the social survey as a basis for town planning is coupled with the ideas of the scientific study of society from the time of Octavia

Hill and the University Settlement movement through the early part of the century. (The University Settlement was founded in Montreal in 1919.) The analysis part usually consisted of exposing all the perceived problems of overcrowding, substandard housing, incidence of disease, crime, and juvenile delinquency, distribution of green space (always negatively correlated with the last), traffic congestion, condition of streets and so on. The plan part consisted of proposing solutions to the problems as they were seen: zoning; minimum property maintenance standards; a parks plan; and street widening schemes. This was then presented to the civic authorities who sometimes acted on parts of it. However, it was during this period of the 1920s that the concept of an overall general comprehensive plan took hold, and early Provincial Planning Acts prescribed the elements to be regulated.

The depression of the 1930s was associated with a swing back to humanist concerns, and in a few circles planning was seen as an important part of restructuring society. However, even the League for Social Reconstruction (1935), while making an eloquent plea for both national planning and town planning, did not divine how local planning could be implemented except through compulsory legislation. It was only after the war and in the aftermath of the orgy of building during reconstruction that the rational model as we understand it today became codified.

From postwar times up to the early 1960s, belief in the power of technocracy reached its peak. This is also the period in which planning borrowed most liberally, often uncritically, from the burgeoning social sciences. Policy science and management theorists started to examine the problem of how decisions are made in a variety of organizations. The model for policy making and planning became the one we know so well: the setting of objectives or goals, data gathering, analysis, elaboration of alternatives, evaluation, choice of one course of action, implementation, review and so on. It is of course never used in the simplistic linear form as noted here. Most planning involves recursive procedures and inputs from many different sources. Robinson (1972) demonstrates a wide range of techniques for dealing with each step in the process.

Although recognizing that the model could be used at any scale or for any sector of activity, its initial overall appeal lay in the promise it held for general comprehensive planning, whether for a municipality or a region. The other approach of these times, that of 'pragmatic' problem solving (e.g. the housing problem or the traffic problem) was known to be imperfect because of the complex web of variables in the urban environment. Today, in fact, most users of the rational model include in the analysis stage an identification of potentials and constraints and from this, special problem identification. This provides an example of how different perspectives can be combined.

In its original formulation the rational model was seen to be value neutral, internally consistent, self-evidently in the public good, and because of the supposedly dispassionate nature of the analysis and projections, providing the 'right' answers to problems. This model, with variations, soon became enshrined in all planning textbooks in some form or another, and with the advent of systems analysis appeared to take on much of the rigour of a more exact science (e.g. Branch 1983, Chadwick 1971, Chapin 1965, McLoughlin 1969).

Even as the rational model was being formalized and incorporated into planners' daily lives, however, it was being questioned, modified, amended, and has come to be seen as an idealistic guide to action which is not attainable but nevertheless useful. Some of the first questioning came from Meyerson (1956), who looked at the problems of long range planning and proposed a system of 'middle range' planning. This was followed by a whole series of studies of the rational method by policy analysts who came to the conclusion that planning is an incremental process (Lindblom 1965), and that because it is not possible for any decision maker to know all the facts relating to a problem, or to be value free, impartial comprehensive decision making is impossible. The new descriptions of planning strategies included 'satisficing' (Simon 1957), 'muddling through' (Lindblom 1959), 'mutual adjustment' (Lindblom 1959), 'disjointed incrementalism' (Braybrooke & Lindblom 1963) and 'mixed scanning' (Etzioni 1967).

These analyses have received much attention by planning theorists (see for example, Faludi 1973b, Friedmann & Hudson 1974). They are very interesting because they describe how decisions are made in the real world. When they are raised to prescriptions for decision making, however, as they have been by some planners, then the endeavour to make real improvement in the urban environment is weakened. These strategies make decision making possible in the inevitable context of imperfect, biased information and competing claims, but do not ameliorate the quality of proscriptive intervention. The most recent manifestation of this approach is typified in the growth of 'mediation' and 'negotiation' techniques to resolve environmental disputes (Susskind 1983). While reaching solutions that conflicting interests can 'live with', the central ideal of optimization is lost. A series of modestly acceptable short-term solutions defuses intractable situations, but in the excitement of conflict resolution, the objective of high standards is diluted.

A further problem with the comprehensive–rational model of planning is that it assumes that there is a common good, and that there can be a societal consensus on goals. Davidoff (1965), recognizing the pluralistic nature of society, and with the disadvantaged mostly in mind, proposed a system of advocacy planning. Here professionals would commit themselves to become the advocates for a specific group and press their claims in this competing world. This was in line with the Alinsky (1971) model of social activism, and was largely the origin of 'store front' planning offices in large municipalities in the late 1960s, a feature that has almost disappeared. Wilson (1983) provides a graphic account of the political hazards in Vancouver.

Advocacy planning represents a conflictual model. It provokes resistance from provincial and local governments, and may disadvantage one group at the expense of others. In the field of co-operative housing, however, it has become institutionalized since the late 1970s in Canada in the form of Community Resource Organizations. Wharf (1980) has described the problems in detail.

An interesting extension of the advocacy paradigm, inspired by Rawls' (1971) *A theory of justice*, has been proposed by Kiernan (1982). Briefly, he proposes a theory of positive discrimination, that is, a code where planners would deliberately favour the disadvantaged, or areas of the city containing low income groups. Whether any municipal government will accept such action remains to be seen. Traditional wisdom has it that municipalities are not agents of redistribution. This is the business of the provincial and federal governments.

Reactions to the rational model and its technocratic accoutrements also came because of its distancing from the people it was supposed to serve. The citizens' groups that sprang up in the early 1960s, for a mixture of reasons, made it abundantly clear that they were not interested in efficiency and optimization, but were interested in democracy, open decision making and neighbourhood integrity. (For a retrospective on a number of Canadian cities, see Roussopoulos 1982.)

This led to another modification of planning theory, the need to make citizen participation an integral part of the planning process. Arnstein (1969) has shown how this can vary in practice from plain manipulation to true participatory democracy. In most jurisdictions it is probably somewhere in between, but at least it has become a part of statutory procedure in most Canadian cities.

In contemporary diagrams of the rational process it will usually be indicated during the goals formulation stage, in the evaluation of alternatives stage and in the preparation of the final schema.

The environmental crisis of the 1960s and the growth of citizen-based environmental protection groups also precipitated changes in planning theory. As the movement gained ground, fanned by activist authors such as Rachel Carson (1962), Reich (1970) and Commoner (1971), planners began to look much more closely at the biophysical environment. In 1969 Ian McHarg published his book exhorting us to *Plan with nature*. In it he demonstrated through a sieve-map technique how all areas unsuited to development because of fragile environments could be identified, mapped and, hopefully, protected. A wave of environmental determinism swept through the profession, and this too was incorporated into the rational model, sharpening up the biophysical survey stage.

The institutional response to the crisis was to pass Environmental Protection Acts, both at the federal and provincial levels. These all came into force around 1972, and essentially demanded that environmental impact assessments be prepared for all major development projects and transportation networks. (Details of the evolution and case studies of the instrumentation of the legislation can be found in Lang & Armour 1980. These are not essential to the present discussion.) Apart from giving planners more work, these initiatives spurred on the development of evaluative techniques. These are usually grouped under the headings of biophysical, social, economic and political. Of these, social impact assessment techniques, with major input from social indicators research and humanist perspectives, have made major advances. Not only primary, but secondary and tertiary impacts can now be unravelled.

Social learning theory also emerged in the early 1970s. It postulated that the gap between technical knowledge and personal knowledge could be bridged by a mutual learning process. Friedmann (1973) proposed a theory of 'transactive' planning, arguing that an interactive approach, with a heavy emphasis on verbal rather than written communication, would result in mutual learning, the closing of the communication gap, and ultimately lead to the creation of a 'societal guidance system'. He had no real prescription for the instrumentation of such a theory, but a planners' meetings with a native people's band council are evocative of this approach.

The crises of human identity occasioned by the technological age were being pinpointed long before citizen participation became an issue (e.g. Ellul 1954). Lewis Mumford published his warning about technics and human development in 1967. (It is of interest that Mumford was a friend of Geddes, and at the time of writing is still living.) The humanist tradition, in the form of the slogan 'people matter more' came to the fore along with the flower children in the late 1960s. The planning response to this was to try to develop a humanist model. Friedmann's transactive theory represents one attempt, but the approach is best exemplified in Lash's (1976) book, *Planning in a human way*. In this account of the author's work as head planner for the Greater Vancouver Regional District, he describes planning as a three-way communication process – graphically labelled a six-sided triangle – between politicians, people and planners. His emphasis is on keeping all methods and channels of dialogue open. The planning process, however, still contained the elements of the rational model, the search for a consensus on goals, alternative options for development and so forth.

The publication of O'Connor's (1973) book on *The fiscal crisis of the state*, Harvey's (1973) *Social justice and the city* and Simmie's (1974) *Citizens in conflict* really marks the beginning of planners' interest in neo-Marxist approaches. The perspectives and interpretations of critical theory vary widely, but in general can be grouped into three: the mechanics of capital accumulation; the role of the state; and class conflict (Dear & Scott 1981). Social and property relations have of course been of concern to planners for generations; the early issues of the *Journal of the Town Planning Institute of Canada* are full of rhetoric decrying the evils of capitalism and equating them with the ills of the working-class society. They were equally convinced that urban society could be reconstructed on a healthy basis through regulation and control, a conclusion still under dispute.

The value of neo-Marxist theory is in its analytic insights into social control, the way in which modes of production control those of reproduction, and the role of the state (once fondly thought to be a neutral referee) in the process of capital accumulation (Wolfe *et al.* 1980). Paris (1982) endeavours to show how planning waxes and wanes, not so much because of its own goodness or badness but according to its value to the state at any given time. As most planners are directly or indirectly employed by the state, or if not, by some private developer, these analyses have given rise to some personal anxiety among members of the profession because they are largely portrayed as lackeys of the system, thus perpetuating inequity (Roweis 1982).

More recent attention has focused on the critical theory of

Habermas, who, among other things, has examined the question of communicative action, of which planning is one example (Albrecht & Lim 1986). He postulates four conditions for communication to operate effectively: (1) that it is true; (2) that it is properly in context (legitimacy); (3) that it is sincerely expressed; and (4) that it is clearly put forward. He argues that if these are accepted, then the social relations of knowledge, consent, trust and attention are reproduced. So far planners have taken these as moral warnings (Forester 1980).

Criticism of the neo-Marxists focuses on the definition of the state, the perceptions of class relationships and absence of any overt manifestations of repression in Canada. The existence of three levels of government, which seem to spend much time in heated dispute and jealously guard their autonomy, stands at odds with the given wisdom of a monolithic state. Similarly, state capitalism, in the form of crown corporations such as Hydro-Quebec or Air Canada, muddy the arguments. Further, in these times of encouraged entrepreneurship, it is often difficult to identify the exploiters and the exploited.

Although the neo-Marxist analyses of the urban condition are very telling and provide powerful insights, they fall squarely into the category of explanatory theory. This group of scholars is still groping to find a way in which their work can be translated into action without the wholesale structural change in Western society that their perspective demands (Beauregard 1980, Forester 1980, 1982). Most radical planners end up with rather timid suggestions for improving forecasting, incremental development, pursuing further research, 'knowing your enemy', and being sure of being aware of the consequences when indulging in any planning activity. They do not present the robust arguments of an earlier generation, for instance Blumenfeld (1967), who has said in many essays that the only way that the conflict between free market forces and human needs will be resolved is through public ownership of all or most of the land to be developed.

PROSPECTS FOR THE FUTURE

It is evident that planning operates under conditions of ambiguity, uncertainty, value conflict, shifting understandings and changing societal mores. Though the search for a general theory of planning such as sought by Ozbekhan (1968) seems to have receded, the search for an adequate standard of ethics is in full swing (Hoch 1984, Howe & Kaufman 1979). This may well represent a response to neo-Marxist and critical analysis. Two other trends seem to be emerging. One is related to the wave of neo-conservatism, and the other toward the movement for self-sufficiency.

The revival of neo-conservatism, as manifested by the election of Brian Mulroney in 1984 (following those of Margaret Thatcher and Ronald Reagan), has incorporated pro-market and anti-bureaucracy stances into local, provincial and national policy debates of all sorts. Planners associated with the public sector are the victims of such attack, partly because they are civil servants and partly because they are seen to stand in the way of entrepreneurial activity. The present style of government, whether local or higher, borrows concepts from the business world such as a strategic planning approach, contracting services out and privatizing others (Savas 1982).

The emphasis on competition seems to deny any idea of redistribution or of distributive justice. Planners who have spent their lives trying to referee, manage and control the inefficiencies, externalities and inequities of the private sector are going to have trouble in reconciling this with an imposed assumption that the self-regulating market is benignly efficient. The description by Crooks (1983) of the criminal irregularities in private garbage collection in Canada and the United States has become well known. Union-busting may also be a reason for privatization, rather than the stated objective of efficiency.

The second trend is towards mutual aid and self-reliance. Goodman (1971), Schumacher (1974) and most recently Morris (1984) have all pointed in this direction for somewhat different reasons: Goodman because of his belief in community, Schumacher to overcome alienation in the workplace, and Morris to counteract economic, energy and environmental problems in the city. The meaning of all this to planning theory is that it will probably take another turn to incorporate the guidance of small groups within its framework. One thing that seems reasonably certain, however, is that it will somehow operate within an again modified rational model.

Alexander (1984) has posed the question, 'After rationality, what?', but this belies the evidence from our legislated provincial planning structures, all of which are based squarely on a positivist conception of the planning process. The most recent provincial Planning Act, that of Quebec (1979), which contains in its preamble the recognition that planning is a political act, which gives municipalities the right to choose which regional municipal counties they would join, and which has mandatory citizen participation built in at three junctures in the process, is set completely in the rationalist mode. Further, the latest Canadian planning textbook (Hodge 1986) directly and unapologetically describes planning as a rational activity. Thus, despite the fact that Kiernan (1982) and many other writers may wish the rational model a speedy death, it is still the basis for planning practice in Canada.

4

Synthesis in human geography: a demonstration of historical materialism

RICHARD HARRIS

INTRODUCTION

Lately, the conception of geography as a synthetic discipline has been gaining ground. According to this view, places are seen as wholes that contain many related parts. Of course the parts can be studied in their own right. This, indeed, has been the project of most academic disciplines, including analytic human and physical geography. But the synthetic view, and according to Cole Harris, the geographer's peculiar habit of mind, consists of 'seeing together the complex of factors that make up the character of places, regions, or landscapes' (Harris 1971, p.162). Many of those who take such a view have concentrated on specific places at particular times, through a more or less modest regional specialization. Others have focused their attention even more narrowly on some specific aspect of a place and time, such as residential differentiation in 19th-century Leeds.[1] Such work may appear similar to that of the analytic researcher. It is distinguished, however, by the concern to identify contextual influences, outcomes and effects, rather than to seek generalizations valid across space and/or time (for example, Sayer 1982a). Always, the synthetic view seeks to elucidate the particular.

As such, it is not new. Following the examples of Carl Sauer and Vidal de la Blache, historical and regional geographers have long recognized the importance of viewing things in context, as parts of a whole. In North America over most of the past 30 years, however, they have constituted a declining minority within the discipline as a whole. Recently, however, the synthetic view has begun to find new support within the fields of urban, economic and social geography. In a manifesto for historical materialists, Harvey (1984, p.4) deplores the fact that geographers have lost their '*raison d'être* as synthesisers of knowledge in its spatial aspect'; Ley (1981) has called for a contextual approach in which the boundedness, as well as the meaning, of human actions is acknowledged; while both Thrift (1983, p.25) and Pred (1984a, b) look forward to a 'new regional geography' in which the complex interpenetration of time and place are brought together. Such ideas are also gaining currency outside the discipline, and Allen Scott has gone so far as to claim that, with 'territorial complexes' as its focus, 'theoretical human geography may emerge as *the* exciting research frontier within the social sciences at large in the 1980s and 1990s' (Scott 1984, p.121). Albeit in strange guise, the synthetic view is making a comeback.

There is no agreement, however, as to how synthesis might best be achieved. The examples offered by Sauer and Vidal de la Blache are in important respects inadequate to present needs. Most people in the developed capitalist societies live in cities. No human geographer interested in the present can afford to ignore the causes, meaning and consequences of that fact. Unfortunately, the works of Sauer and Vidal de la Blache do little to illuminate the issue, as they deal for the most part with rural life in the past (Buttimer 1978, Williams 1983). Such biases may still be found among historical and regional geographers. There is a further limitation. It is now widely believed that the study of human geography raises theoretical questions as to the nature of, and relationship between, the individual and society. These questions have concerned, indeed preoccupied, philosophers and social scientists in the 20th century, some geographers included. They inform recent exchanges between Marxists and humanists over the nature and relative importance of human agency and social structure (Chouinard & Fincher 1983, Duncan & Ley 1982, 1983, Gregory 1981, Ley 1982). But they were not issues which either Sauer or Vidal de la Blache chose to discuss. The great strength of these writers lay in their practice of, rather than in their ability to theorize about, the synthetic method. In his programmatic statements, for example, Sauer cheerfully contradicted himself (Williams 1983, p.2). A reticence in the face of theory, amounting in some to distrust, is to be found today in the writings of many historical geographers. Even Cole Harris, who has been more willing than most to make general statements about his field, prefers talk of a 'habit of mind' rather than a method, emphasizing the essentially elusive quality of historical judgements (Harris 1971, 1978). I do not see anything about historical geography which is inherently incompatible with theoretical debate. It is a fact, however, that historical and regional geographers have skirted many of the issues that trouble human geographers today. This, more than their concern for rural ways and past life, will continue to limit the relevance of the message of Sauer and Vidal de la Blache.

But the more modern alternatives have not won a consensus either. Strong claims have been made for two views, humanism and Marxism, and proponents of each have criticized one another.[2] Humanists have emphasized the ways people find meaning through social action. They have been criticized, I believe correctly, for failing to offer an adequate theory of the social constraints upon human action (Gregory 1981). Marxists, on the other hand, have a well-developed theory of social structure. They believe that material life determines – in the sense of setting limits to – the prevailing character of social ideologies and political organization (Williams 1977, p.85). Some, and in particular some structuralists, have gone further in arguing that structures determine – in the narrower sense – what people think and do. In so doing, the latter have been rebuked for excessive abstraction and economic determinism (Duncan & Ley 1982). Jackson & Smith, for example, have noted the 'difficulties' experienced by Marxists in 'translating theoretical categories into active groups of men and women' (Jackson & Smith 1984, p.196). Reasonable people on both sides have conceded these points, but have gone on to claim that any failures are those of imagination and ingenuity, and are not inherent in the approaches themselves. Thus, in precisely these terms, S. Smith (1984) has recently defended a pragmatist version of humanism. Similarly, conceding that 'the more structuralist versions of Marxism have. . .an attenuated conception. . .of human agency', Gregory (1981, p.16) insists that the fault is not intrinsic to Marxism. Nevertheless, he does suggest that this approach might benefit from 'reconstruction' in terms of the concepts developed by Giddens in his 'theory of structuration' (Giddens 1979). This, Gregory believes, will make Marxism more capable of 'comprehending intentionality and consciousness' (Gregory 1981, p.16).

I believe that Gregory's assessments of humanism and Marxism are sound, but that reconstruction of the latter is unnecessary. The argument could be made in theoretical terms. Indeed, at least in a tentative form it already has.[3] But in recent years human geography has been overburdened with theoretical statements while being starved of theoretically informed empirical research. In my view there is now a greater need for the latter than the former. The purpose of this chapter, then, is not merely to argue but to demonstrate that an unreconstructed Marxism can accommodate the role played by people in the shaping of events, and without sacrificing the concept of determination. This demonstration takes the form of an interpretation of the role played by a group of 'New Left' organizers in Kingston (Ontario) city politics between 1965 and 1970.

MARXIST SYNTHESIS IN PRACTICE

To rest a general argument on a case study of such limited scope might seem irresponsible. In fact it makes a good deal of sense. Marxists have often prided themselves on their ability to comprehend the big picture. Even their critics have acknowledged a strength in this area, reserving criticism for the Marxists' treatment of smaller events, of particular places and lives (for example, Sartre 1963). Within geography, Harvey concedes that 'Marxists. . .have had a hard time. . .evolving a sensitivity to place and milieu' (Harvey 1984, p.9). By examining local politics in a small Canadian city it is possible to meet this concern head on.

My interpretation of the Kingston New Left places a great deal of emphasis upon the concept, and reality, of class. In the Marxist view, class is viewed as an objective economic relation between groups of people and also between people and their means of production. In these terms, and following the work of Carchedi, four major classes are identified: owners and managers, the self-employed, the middle class and the working class (Carchedi 1977). As those who have written about dual labour markets have emphasized, however, the North American working class is divided into two fractions: a primary sector containing workers who are unionized and relatively well paid, and a secondary sector comprising those who are unorganized, poorly paid and on the fringe of the labour market (Doeringer & Piore 1971; Edwards *et al.* 1975, Smith 1976). In these terms, the New Left was a social movement composed predominantly of middle-class youth. To the extent that it became involved in grass-roots political organizing it concentrated its energies on secondary sector workers, the working and welfare poor. I argue that in Kingston, as elsewhere, the class composition and base of the movement set limits to its achievement. As Marxists such as Thompson (1968) have often insisted, however, class is not only an objective relation but also a subjective appraisal. That makes it a peculiarly rich concept, one that is correspondingly difficult to apply. In Kingston, many of the people that the New Left organized – and some in the New Left itself – did not think of their situation in class terms. This was important. But in the political alignments and eventual polarization that the actions of the New Left produced, the underlying and divisive influence of class may be seen. It is not necessary for the importance of class to be acknowledged in order that it be felt. It is this characteristically materialist conception of class that shapes the following account.

If class was the most important determinant of the role of the New Left, it was by no means the only one. To an unusual degree the movement in Kingston was able to mobilize the support of a broad working- and middle-class base, and with this support to win concessions from the local state. To understand why, it is necessary to consider as a whole the historical and geographical milieu of Kingston in the 1960s. Among other factors, the residential geography of the place, local housing conditions, and the leadership of committed organizers all played a vital part in shaping the character and impact of the New Left. In emphasizing the importance of such contextual influences and effects, this interpretation of the character, origins and effects of the Kingston New Left exemplifies the synthetic approach.

CIRCUMSTANTIAL ORIGINS

The 'New Left' that developed in the aftermath of the Cold War is not easy to define. In a general way this mainly middle-class movement differed from existing communist, socialist and social democratic parties in that it sought to bring about radical change through mass-based political organizing.[4] Ideologically, however, it was diverse. This was particularly true in the early years, when radical populism, anarchism, socialism and religious moralism each contended for dominance. Moreover, its character varied from place to place while always being in flux. The Canadian movement was shaped by the preoccupations and tactics of the American New Left, these being grafted on to disarmament concerns more typically British (Laxer 1970, Levitt 1984, pp.47–8, 158–63, Westhues 1975). As the Canadian movement developed, regional differences emerged. It has been observed that the American populist tradition and the community organizing strategy of the American Students for a Democratic Society (SDS) appear to have had their greatest impact in Ontario (Laxer 1970, p.185); certainly Kingston's movement was shaped by such influences, especially in the early days. Moreover the movement was never still. In English Canada, the moral attack on disarmament which occupied the Combined Universities Campaign for Nuclear Disarmament (CUCND) from 1960 to 1964 gave way in 1965 to the broader concerns of the Student Union for Peace Action (SUPA). From 1966, as the movement confined itself to campus, it became increasingly Marxist and sectarian. In 1967 SUPA dissolved into fragments. With the major exception of the women's movement, these soon dropped the New Left's characteristic emphasis on participatory democracy and lost contact with popularly felt concerns.

In Kingston, propitious circumstances allowed the local New Left to continue organizing long after the national movement had fallen apart.[5] In terms of its social and class composition, Kingston offered the New Left much potential. University students, the base of the New Left, made up a high proportion of the city's population, at least during the school year. In the 1960s, the student population of Queen's University grew to 10,000 while the population of the city as a whole remained steady at six times that amount. The size of Queen's in relation to the city helped to ensure that the steady stream of volunteers for community organizing work would have a significant political effect. Organizers were to get valuable assistance from a 'Waffle' group that was also based on campus. The Waffle developed in the late 1960s as a left caucus of the New Democratic Party (NDP), a social-democratic party roughly equivalent to Britain's SDP (Hackett 1980). Locally, a Waffle group gave support to the community organizing efforts of the New Left. By the beginning of 1968 it had gained control over the NDP constituency association, which then helped to finance the Community Information Services, a service and advocacy centre for low-income people and tenants. Here again, through its effect on the local NDP, the presence of Queen's gave the New Left a boost.

Of at least equal importance was the fact that the working and welfare poor, the New Left's main political constituency, made up a relatively high proportion of the population. Kingston was (and is) a reception area for the rural migrant poor of Eastern Ontario, and throughout the 1960s its rate of welfare dependency was about double that of the province as a whole. The steady growth of Queen's put pressure on the local rental market. For a city of its size, Kingston had among the highest rents in the province, and by 1970 a survey revealed that vacancy rates in the downtown area had dropped to almost zero (Canada Central Mortgage and Housing Corporation 1969). Much of the housing stock in the inner areas was almost a century old, and living conditions in some dwellings were appalling. Housing, then, was an obvious issue for organizing, one on which middle-class students and poor people could find common ground.

The task of organizing was made easier by the geography of the place. Size alone was important. Elsewhere, the New Left became sectarian in 1967. Kingston, however, was too small to support factional divisions, these requiring higher thresholds than the available minorities of one. Since most places in the city could be reached on foot, it was relatively easy for organizers to get people to demonstrations and public meetings. Social segregation also helped. Kingston was divided into three very distinct areas: the middle-class West End, the working-class North End and a very mixed neighbourhood downtown, known as Sydenham Ward (Fig. 4.1, *cf.* Harris 1984, pp.459–63). St Lawrence Ward, in the inner North End, contained a particularly high proportion of low-income tenants. The concentration of the New Left's constituency in this area made organizing relatively easy: this was true both in the summer of 1965 and again with the formation of the Association for Tenants' Action, Kingston (ATAK) in 1968. When ATAK's president ran successfully in the 1968 civic election, she chose St Lawrence as the ward where she was most likely to succeed. If the poor had not been concentrated in this area it is unlikely that ATAK would have secured a voice on City Council. In sum, geographical as well as social circumstances conspired to favour the efforts of the New Left in Kingston.

ACTIONS

The advantages to organizing in Kingston were not apparent to the group of students who set up the Kingston Community Project in the summer of 1965. Outsiders who knew nothing about the city, they set out to 'start an organisation of poor people to solve their own problems' (Kingston Community Project 1965a). Having defined a project area in the inner North End, they went knocking on doors. They found plenty of issues, including high rents, poor housing conditions and unsafe streets, but had little success in getting support for reform. Area residents had little sense of community (Kingston Community Project

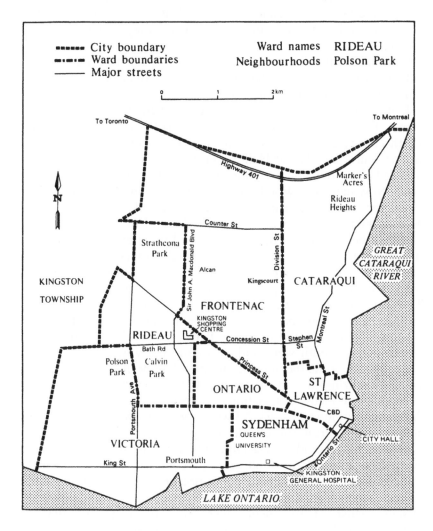

Figure 4.1 Kingston township

1965b). Indeed, as is often the case, there was some active hostility between the primary sector working class and the welfare poor. Being middle-class outsiders, the students were viewed with suspicion and that autumn, discouraged, they went back to school.

Fortunately, the director of the Project was able to bring in two full-time organizers to replace them. Using contacts made in the summer, Joan Newman and Myrna Wood seriously set about the task of organizing. Their first successes were with North End youth. For many years the North End had been the base for a youth culture of working-class dissent. On occasion this erupted into violence, as it had at Hallowe'en in 1962, when a mob of 400 stormed Kingston police station, to be beaten off only with tear gas.[6] Newman and Wood made contacts, and encouraged the formation of a youth-run coffee-house, the 'Needle's Eye'. The 'Eye' attracted many people, and by the end of the summer of 1966 had been hounded out of existence by the local fire marshal and the police. Other clubs were formed, however, and out of these there grew an increasingly articulate youth counter-culture. After a couple of attempts, an ambitious alternative newspaper, *This Paper Belongs to the People*, was set up in 1969 in order to provide 'a different perspective and a better analysis' than the existing media.[7] With the New Left as a catalyst, Kingston's working-class youth were finding a common voice.

At the same time the housing situation had been getting steadily worse. Realizing this, in the spring of 1967, activists began to try to organize residents' associations to press for the rehabilitation of the inner city housing stock.[8] Two associations were formed, but neither was a great success. It was partly out of such failures, however, that the idea of a community information centre was born. Set up early in 1968, and in contrast to the residents' associations, the Community Information Service was an immediate success. Apart from service work, CIS became a centre of political dissent. Most of its clients were tenants, many of whose complaints and concerns arose out of their housing situation. As a result, organizers soon hit upon the idea of starting a tenants' association. By September ATAK boasted a membership of over 800. Apart from the problems of slum housing, one of the more important reasons for the success of ATAK was the relatively high tenancy rate in the city. Less than half of all Kingston households were owner-occupiers at that time. This was about 20 percentage points lower than the rate for other Ontario towns of comparable size. Tenancy rates varied little between the working and middle classes (Harris 1984, p.486). As a result, those who joined ATAK in the fall of 1968 included a substantial number of middle-class people, as well as students and the working class. The membership drive met with so much success that the decision was made to run three candidates in the upcoming civic election. Two were elected, Newman in St Lawrence and John Meister, a shop steward, on a joint ATAK-NDP slate in Ontario Ward (see Fig. 4.1). After more than three years of organizing the New Left had finally mobilized the working and welfare poor.

Overnight, Meister and Newman changed the tone and content of debate on City Council.[9] Hitherto Kingston politics had been run by local businessmen, and largely from the Chamber of Commerce. Now this group was challenged at every turn: zoning, housing and welfare policy, the location of parks, all were dragged out for public scrutiny. The local franchise, which allowed people to vote in any ward in which they owned property, was singled out for criticism. ATAK kept in touch with its North End roots. The information centre continued after NDP support was withdrawn in the autumn of 1968 and became a storefront for ATAK. Other tenants' groups formed and some affiliated as 'locals' to the main association. A rent strike was launched against a notorious local slumlord. Support pickets were organized for the electrical workers at the locomotive works. In 1968, and the early months of 1969, ATAK was at the centre of an active culture of extra-parliamentary dissent, reflected in, and articulated through, the pages of *This Paper*.

With its own press, popular support and representatives on Council, ATAK began to polarize Kingston politics.[10] In the civic election campaign of 1968 ATAK had received support from the NDP, local unions and the Labour Council, while goading landlords into organized opposition. Through 1969 it continued to work closely with the Waffle-dominated NDP, and indeed many people were members of both organizations. In the spring of 1970 a developer unveiled plans for high-rises on the waterfront. It provoked immediate opposition from ATAK, the NDP and a local ratepayers' group. The plans were withdrawn. At the same time ATAK was pressing its demand for rent controls. The issue had been debated many times. Uncertainty remained as to whether the City had the legal power to enact such legislation, but ATAK, with support from the NDP, the Social Planning Council, the Labour Council and student groups, was insisting that Kingston make itself a test case. On this issue, students and the working class, along with a substantial fraction of the middle class, found themselves ranged against the small businessmen, landlords and associated professionals who for the most part controlled Council. The issue came to a final vote in August 1970, when a motion to this effect was soundly defeated. As spectators disrupted the proceedings, Council was forced to adjourn, for the first time in Kingston's history.

This defeat marked a turning of the tide. Having thrown itself into the battle on Council, ATAK had begun to lose touch with its constituency. Partly as a result of organizers' efforts, several hundred public housing units were on stream, holding out a promise that the housing situation might improve. The FLQ crisis in Quebec that fall helped to give radicalism a bad name.[11] For a combination of reasons, then, when four candidates ran on a joint NDP-ATAK ticket in the December civic election, all were defeated. It was the end of an era.

ACHIEVEMENTS

Now that the smoke has cleared it does not seem that very much was achieved. To be sure, the New Left had succeeded in mobilizing some of the poor. This was no mean feat, given the historic conservatism and political alienation of that group, in Kingston as elsewhere (Cloward & Piven 1979). Indeed, with this base, they had helped to provoke a polarization in Kingston politics, which saw the working class and part of the middle class ranged against landlords and small businessmen. Not everyone who took sides viewed the situation in class terms, but it is striking how sharply the class lines were drawn. Only the middle class was divided, salaried professionals (especially those at Queen's) being more inclined to favour the NDP-ATAK coalition, while those associated with the local real estate market lined up on the other side. The effect on the conditions and relations of life in Kingston, however, was small. The New Left helped force City Council to put up a couple of 'stop' signs on one of the more dangerous North End streets, and to establish a playground in the same area. Better housing by-laws were passed and enforced, and slum landlords coerced into cleaning up their act, mainly as a result of the pressures brought to bear through CIS and ATAK. City government was made a little more democratic and responsive to popular demands, for example through the institutionalization of public meetings to deal with planning issues. The New Left played an important part in preventing unsightly waterfront redevelopment and in getting public housing built in the far north of the North End. In this way they helped to change the geography of the city. But the consequences were neither entirely fortunate nor particularly long-lasting. When built, the isolated public housing ghetto created as many problems as it solved, while, more recently, high-rise buildings have been put up on the waterfront which are no more suitable to the site than those proposed in 1970. Overall the achievements, if not negligible, were certainly modest.

The reasons are clear. As New Left organizers themselves realized from the very beginning, the poor are powerless. Unorganized and often unemployed, they can impose few sanctions. They cannot stop production by striking, because most are unemployed or unorganized. In the sphere of reproduction, rent strikes, boycotts and the physical disruption of public meetings are the main options. All were attempted in Kingston. Unless backed by massive and widespread support, however, such tactics are unlikely to win significant concessions (Cloward & Piven 1979). This is particularly true as long as they are directed at local government. In Canada, more than in the United States, local governments are weak (Higgins 1977, Plunkett 1968). Such responsibilities as they possess have been delegated to them by the provinces and, in principle, can be revoked. They can make few, and for the most part insignificant, concessions, and even these can be reversed by upper levels of government. Moreover, the prospects for changing state policy at higher levels from a Kingston base were small. The movement in other cities was not as strong. This might have mattered less if Kingston had loomed as large on the provincial or national scene as did, say, Toronto. But of course it did not. The size of the city, then, coupled with the movement's concern to organize the poor to demand changes in local government policy, set narrow limits to the possible achievement of the local New Left.

PEOPLE MAKING HISTORY

The high level of political support coupled with the modest achievements of the Kingston New Left may be understood in terms of the circumstances in which organizers worked, but not wholly so. The local New Left was created by a group of people determined to bring about change. A full analysis of the movement must examine the active role played by people in taking, and shaping, the opportunities provided by their social, geographical and historical milieu.

To some extent this role may be understood in terms of the actions of particular individuals, those whom historians call Great

Men.[12] Even so, a variation on this theme must be introduced, for in Kingston most of the great men were, in fact, women. Over the period as a whole the most important person, of either sex, was Joan Newman. Committed, able and articulate, she devoted herself for five years to the task of community organizing in Kingston. Her persistence helped to overcome the suspicion with which middle-class organizers were first greeted. She played a major role in virtually all of the most important organizations: in the original coffee-house, in CIS and ATAK, and finally on City Council. And yet a number of others played important parts.[13]

Giving due weight to the leaders, however, does not exhaust the role played by human agency. It is perhaps obvious that, without popular support, organizations like CIS and ATAK would have counted for little. In that sense, every member, every person on a picket line, every voter, played a part. Less obvious is the fact that they gave the movement a life of its own. This often took the 'leaders' by surprise. From interviews that I conducted, it became clear that none of the most active organizers anticipated the success of the first coffee-house, of CIS or ATAK.[14] Like the residents' associations, these were experiments; unlike them, they worked. In one sense, then, the organizers were merely catalysts that allowed the real agents of change, middle-class students, working-class youth and the poor, to have effect.

But in one respect the metaphor of agent and catalyst is inadequate. The catalysts themselves were altered. Newman claims to have learned her politics 'on the street' and indeed the views of most activists had been shaped by the experience of collective struggle. The extent of the change can be gauged by contrasting the stated purposes of the New Left at the beginning and end of its involvement in Kingston politics. In 1965 the movement was trying to help the poor to help themselves, and without much idea of how to do it. By 1970, knowing far more about the problems and potentialities of organizing in Kingston, they wanted, in Newman's words, 'to fight for. . .basic change where it matters: in the homes, streets and workplaces of this city'.[15] The achievements were much more modest than that, but in trying to make history they did remake themselves.

DISCUSSION AND IMPLICATIONS

This account of the New Left in Kingston is synthetic in that its primary concern is to understand local political developments in their context, rather than to establish generalizations about similar movements in other cities. This context, the social and geographical circumstances in which the New Left acted, was in part local. The class structure, the size and growth of Queen's and the size and geography of the city itself, provided a unique (and rather propitious) set of opportunities and constraints. In part, however, the context was national and indeed international. The unfreezing of the Cold War and the perceived failures of existing parties of the Left defined the distinctive political character of the movement. But it is clear that people were not merely agents of class interests, passively responding to some inscrutable historical and geographical necessity. Individually and collectively they responded to, and sought to change, their situation. This synthetic account, then, assumes and demonstrates that people make their own history, albeit under circumstances that are not of their own choosing.

In a general way this formula, borrowed from Marx, is similar to those which are already used by some geographers and historians. To be sure, it is broader in scope than that implied by the terms 'agency and structure', for the latter implicitly excludes the environment. But it does come close to the framework of 'character and circumstance' used by many historians;[16] it bears some similarity to Sauer's treatment of culture and environment and to Vidal's *genre de vie* and *milieu*; it is entirely consistent with the way some humanist geographers discuss action and context. In that respect, humanists, historical geographers and for that matter historians might all provide a broadly similar interpretation to the one offered here.

In other respects, however, their accounts would surely differ from this one. The meaning, as well as the weight, given to 'class' in this interpretation of the New Left is distinctive. The notion that differences in class interests lay behind the polarization of Kingston politics, even though many of those involved would have denied it, is materialist; as such it would not recommend itself to many humanistic or historical geographers. Moreover, the particular classes identified in this account are not identical to those defined by other materialist approaches, for example that developed by Max Weber. It is important to recognize that Marxism, humanism and historical geography share the synthetic approach, for this effectively highlights the common basis from which discussion can proceed. But it is surely misguided to propose, as Gregory and others have done, that these approaches can be reconciled. They share a common denominator but they remain different quantities.

The Marxist approach has much to offer geographers. Its synthetic approach is quite compatible with the habit of mind of the historical and regional geographic traditions. It has an unusually well-developed theory of both social structure and change that sets it apart from recent humanist thought. This theory embodies a materialist conception of determination which most geographers, steeped as we are in environmental modes of thinking, should find congenial. Moreover, as I have tried to show in this case study of Kingston, the theory is not as inflexible as many of its critics have claimed. Indeed it is quite capable of giving due weight to the capacity of people to make their own history, and hence their own geography.

Some will object that the Marxist formula is vague and unhelpful because it does not tell us exactly how much power people have to change the circumstances of their lives. But this arises from a misconception. The amount of power people have depends on the place, the time and the people in question. Those who formed the New Left in Kingston, and those whom the New Left mobilized, were relatively powerless. In so far as such judgements are possible, we might say that they remade themselves more than they made history. But we cannot generalize from such a case, for their situation was unique. That, after all, is a major assumption of the synthetic approach. Others may object that, vagueness aside, the formula contains a contradiction: it asserts the simultaneous existence of necessity and choice. That is true, but the contradiction is not merely a difficulty with the theory. It is inherent in the world as we know it. As individuals, groups and political collectivities we continually struggle to better our condition. In doing so we define our purposes against the circumstances that we inherit, while facing constraints and sometimes opposition. We are free and unfree. To understand a paradoxical condition we need a paradoxical method. Far from being a weakness, then, the vagueness and contradiction of historical materialism is its greatest strength. It is a method of analysis, not the pat answer to specific questions. In that sense, it has no magic properties. It is simply the best method we have.

ACKNOWLEDGMENT

I would like to thank Suzanne MacKenzie, Ted Relph and Carol Town for their valuable comments on an earlier draft.

NOTES

1 See Ward (1980). Ward has also used his Leeds research as the basis for speculation about the degree of segregation in English industrial cities in the 19th century. This is a type of generalization which is carefully circumscribed in both historical and geographical terms. Such generalizations have been reconciled with the synthetic view by Cole Harris (1971, p.164), who defines them as 'generic studies' within specific 'cultural regions'. The line between studies of this kind and the generalizations of the analytic tradition is a fine one.

2 In fact, as Marxism is a type of humanism, the recent debate might more properly be defined as being between Marxism and other types of humanism (cf. Ley & Samuels 1978, p.3).

3 See, for example, Sayer's discussion of structural Marxism and Giddens' theory of 'structuration' (Sayer 1982a, p.81).

4 There is a very substantial body of literature, of very uneven quality, concerning the New Left. I have found Levitt's (1984) study of the movement in the US, Canada and West Germany to be the most valuable single work, both for its international perspective and for its analysis of the New Left's social base. Breines (1982) has given us the best overview of the US New Left's community organizing activity.

5 A fuller account of the activities of the New Left in Kingston, together with more extensive documentation, may be found in Harris (1985, 1988).

6 'Charge eleven following riot', front page headline, *Kingston Whig-Standard*, 1 November 1962. On the coffee clubs, see Newman & Crossfield (1968).

7 Editorial statement of intent, July 1969, *This Paper Belongs to the People* **1** (1), 2.

8 The statements in this paragraph are based on Newman & Crossfield (1968) and a review of the *Kingston Whig-Standard*, the local daily newspaper, 1961–1976.

9 Statement based on a review of the minutes of Kingston City Council, and all Council sub-committees, 1961–1976.

10 For a more detailed account see Harris (1985, 1988).

11 This crisis was precipitated by the *Front de Libération Québecois* (FLQ), an extreme separatist group which, on 5 October 1970, abducted the British Trade Commissioner in Montreal and, five days later, the Quebec Labour Minister. Prime Minister Trudeau declared a state of national emergency by invoking the provisions of the War Measures Act. At the time this action was very popular, although in retrospect its wisdom and necessity have been seriously called in question (Postgate & McRoberts 1980).

12 The case for viewing history in terms of the actions and achievements of Great Men was persuasively articulated by Thomas Carlyle (1840). My own thinking on this issue follows closely the brief, but incisive, observations made by Carr (1964, pp.52–5).

13 In the NDP, for example, John Smart did vital work behind the scenes, while Jim Laxer, more concerned with provincial and federal than with local politics, gave the Waffle group charismatic leadership. In a different way, Bobbi Spark did important work with welfare recipients. She had a peculiar legitimacy, being a welfare mother with five children in public housing. Indeed, one of the features of the movement in Kingston was that it attracted working-class people who themselves took on active leadership roles. Unionists like John Meister, and young men like Dennis Crossfield, gave the movement a depth and a legitimacy without which organizers like Joan Newman, however committed, would have been left whistling in the dark.

14 The following people, all politically active in Kingston within the period of study, were interviewed for this study: Ron Baxter, Dennis Crossfield, Dennis McDermott, John Meister, Logan Murray, Joan Newman, John Smart, George Rawlyk, Bobbi Spark, Sarah Spinks, Bronwen Wallace. The 'Big Chill' factor notwithstanding, none of those who were active in the Kingston New Left expressed cynicism about the aims of their activity in the 1960s. Indeed the majority have remained politically active on the left, either in the women's movement or in community and union organizing.

15 Reported in 'Replacement not likely for rest of term', *Kingston Whig-Standard*, 17 September 1970, p.1.

16 In this connection it is interesting that, unprompted, George Rawlyk, an historian active in the Kingston NDP in the late 1960s, made precisely this distinction when discussing with me the role of the local New Left.

SECTION II
Methods

Introduction

The chapters in this section explore some of the controversies which have been raised in our attempts to use historical materialism and humanism to answer questions about environmental relations. They are, in a sense, more reflective than the chapters in the previous section, and also more reflexive, engaging directly with the internal logic of the paradigms, their theory, method and identification of the essential questions that intersect with the reality we purport to study.

The first two, by Pratt and by Foot, Rigby & Webber, examine the basis for quantitative techniques and measurement. Pratt's objective is to 'review several of the central criticisms of positivism advanced through the historical materialist and humanist perspectives, and to consider the extent to which the use of statistical inference and quantification is susceptible to them'. She concludes that 'the use of statistical inference does not necessarily imply the reduction of theory to empirical regularities or the ontological assumption of atomistic monism. It need not necessarily lead to impoverished theoretical categories or a reductive, mechanistic view of human nature'. But she continues,

> the value of the criticisms outlined by historical materialist and humanist geographers is that they emphasize the distance between quantitative research appropriate to these perspectives and that conducted within a positivist philosophical framework. They draw attention to the fact that statistical analysis cannot subsume theoretical development, that theoretically uninformed quantitative research is of little interest and of little use in explaining social processes.

In other words, such criticisms serve to pare quantitative research down to size, to the status of a technique for organizing one type of information, that must necessarily be complemented by a more relational, contextual understanding, as well as more abstract theoretical development.

Foot, Rigby & Webber place their chapter within the discussion of how 'Marxist theory is evaluated by reference to experience. . .

actual historical events'. Their aim is to show 'how abstract laws and concrete events may be linked to provide an evaluation of theory'. This chapter illustrates, using the concept of the tendency for the rate of profit to fall in the concrete cases of the Canadian economy as a whole and the US steel industry in particular, the use of some of the techniques which Pratt discusses. It also specifies the dialectical relation of abstract and concrete and the historical relation of agency and structure. They specify a realist mode of evaluating theory, and explain how this differs from non-realist modes.

Trevor Barnes shares with the previous authors an explicit interest in the dualism of structure and agency in economic geography, and in the use of realism as a basis for reconciliation of structure and agency. He argues that this dualism results from the adoption of one of two 'architectonic' economic theories: the labour theory of value adopted by historical materialists; or utility theory, the basis of neoclassical economics. Claiming that the former 'implies a structural account of economic behaviour' and the latter 'leads to a voluntaristic account', he suggests the adoption of a contextual position, as illustrated in the work of Piero Sraffa.

The relation between agency and structure is also central in the next two chapters, by Kobayashi and Relph, who examine the concept of landscape. Both give landscape an important role not only in disciplinary discourse but also in the reconciliation of agency and structure. Kobayashi is very explicit about this, utilizing and extending concepts of Jean-Paul Sartre to address geographers' concerns with that 'range of structures that define the quality of the human being/environment relationship'. Relph is concerned to 'clarify and describe the use of. . .responsive, phenomenological or artistic methods. . .directed to the study of landscapes'. He argues that landscape study is not amenable to 'scientific and systematic methods', nor is it conducive to 'digging beneath the surface' to seek causal explanation, but that landscape can 'be an entry point for addressing social and political issues, because they are as much a source of information about social conditions as documents, interviews or censuses'.

Both chapters make explicit the social concerns which are implied in the previous three. Kobayashi suggests that 'we may envision another form of landscape in which ideology is expressed as common commitment, shared concerns, the fusion of praxis to create congruence of structure and the agency of the individual'. In some respects this echoes the recent concerns of some historical materialists, such as Castells, with the formation of alliances between groups on the basis of their living space, or of feminists, on the basis of the tradition of home and community based organization among women. In Relph's more immediate terms,

> writing and teaching about landscape can promote independence in thinking and judgement, can help to develop an understanding of the complex and open-ended processes that characterize all human-environmental relationships, and perhaps, in the very long run and through an emphasis on responsiveness and responsibility, they might contribute to the discovery and acceptance of ways of living on the earth which are less technically violent and more sustainable than the ones which now prevail.

5

Quantitative techniques and humanistic–historical materialist perspectives

GERALDINE PRATT

INTRODUCTION

Referring to the development of humanistic, Marxist and 'critical' perspectives by geographers through the 1970s, Szymanski & Agnew have written that:

> All have been long on asserting what ought to be done and have tended to focus on programmatic statements rather than actual research practice. With a few notable exceptions those who enjoy their role as philosopher have proved impotent in demonstrating the substantive or theoretical implications of their message (Szymanski & Agnew 1981, p.66).

Although premature as a serious appraisal of the relevance of these perspectives to empirical research, this statement points to the necessity of integrating these philosophical programmes with more specific methodological precepts, as well as the scarcity of these links. Integration seems even more important given the contemporary enthusiasm for empirical study among Marxists and humanistic geographers and also the oddly familiar quality of some recent empirical research conducted by Marxists and humanists. This familiarity could be seen as odd because it is reminiscent of the quantitative research criticized so effectively from these two perspectives as positivist.

For example, this strong sense of *déjà vu* is elicited by the work of the Marxist sociologists, Wright and Perrone. They state that: 'It is our hope (to) open up the possibility of bringing Marxist categories into the heart of quantitative research on social inequality as well as making quantitative research seem more relevant to Marxist social scientists (Wright & Perrone 1977, p.54). To this end, they present 'a preliminary *operationalization* of the Marxist criteria for class position for use in quantitative research' and examine 'an application of this *operationalization* in the study of income inequality' (1977, pp.32–3, my emphasis), complete with multiple regression equations and levels of statistical significance. That other Marxists have been receptive to this mode of enquiry is indicated by Patrick Dunleavy's statement that the work of Wright & Perrone is 'the best Marxist empirical work on social class boundaries' (Dunleavy 1979, p.4). Dunleavy himself is conducting sophisticated quantitative analyses of political attitudes by housing consumption sectors. Among those espousing a humanistic perspective, David Ley has recently employed attitudinal scales, non-parametric statistical techniques and statistical inference in assessing attitudes toward office relocation (Ley 1985b, see also Ley 1984). In my own work on the links between home-ownership and political values, I have made substantial use of attitudinal measures and non-parametric statistical techniques, while accepting the weight of the humanist critique of positivism (Pratt 1984).

That something is amiss in this embracing of quantitative technique by historical materialists is suggested by Andrew Sayer's remark that: 'The methodological principles of statistical inference *presuppose* the fatally-flawed empiricist situations of atomism and regularity generalisation' (Sayer 1978, p.53). Sayer's claim is that there is a fundamental mismatch between the use of statistical inference and the philosophical tenets of realism, which he takes to underlie historical materialism.[1] The emphasis on understanding the meaning of social phenomena has also prompted humanistic geographers to argue that: 'Methodologically, this requires a move from the principles of statistical inference based on representative random samples, to those of "logical inference" based on unique or idiosyncratic case studies' (Jackson & Smith 1984; also see S. Smith 1984).

The purpose of this chapter is to review several of the central criticisms of positivism (or empiricism) advanced through the historical materialist and humanistic perspectives, and to consider the extent to which the use of statistical inference and quantification is susceptible to them. Though Andrew Sayer is certainly not the only historical materialist to criticize the type of empirical work conducted by positivists (for summaries see Bhaskar 1979, Hindess 1977, Keat & Urry 1975), his criticisms stand out in geography, both because he has been extremely articulate about the specific points of incompatibility between statistical inference and historical materialism and because he has emphasized the necessity for empirical research within a Marxist perspective. Therefore, special attention will be directed toward his detailed criticisms. As noted above, Sayer has modified his position somewhat so that it is compatible with the arguments presented here. His resolutions of the links between a realist philosophy and statistical inference are, however, developed somewhat differently from those developed here.

Altogether four criticisms of positivism will be evaluated in terms of their implications for quantitative research. These include: the critique of theory as empirical regularity and the rejection of an atomistic ontology, both criticisms coming from the historical materialist perspective; a realization that facts are interpreted and, finally, a rejection of the reduction of human-ness to physical properties and hence a neglect of meaning, human intentionality and choice, a criticism common to both perspectives.[2] The arguments in this paper consider the implications of these philosophical criticisms for the *methods* appropriate to historical materialist and humanist perspectives, namely the logical consistency of using quantitative techniques if one accepts these criticisms. In other words, some philosophical groundwork is assumed and this chapter focuses more exclusively on specific methodological issues which follow from preceding philosophical discussions. It will be argued that quantitative techniques are not vulnerable to these more basic philosophical criticisms and that a distinction should be maintained between the logical underpinnings of these techniques and their use by positivist geographers.[3]

EMPIRICAL REGULARITY

From the historical materialist perspective, the relationship between theory and empirical research is generally taken to be defined by a realist philosophy of science (Bhaskar 1979, Keat & Urry 1975, A. Sayer 1984, D. Sayer 1979). A distinguishing feature of the realist position is the rejection of a Humean definition of causality as the constant conjunction of events. For Humeans, causation is equivalent to regular succession; if event B follows event A, we say that A caused B. Humeans do not assume the logical necessity of this temporal relationship. What is at issue here is that nothing is observed except that B follows A. Realists reject this conception of causality in favour of one which admits of causes as mechanisms, such that properties of A, when stimulated, produce B. For example, from the realist perspective we need to say more about the fact that the match lights after the match is struck than that the former follows the latter. We need to describe the causal mechanisms involved. An adequate causal explanation of this phenomenon consists of a description of *how* striking the match causes the match to light and this is in part explained by describing the nature of the match and the surface against which it is struck. Causes are conceptualized as necessary ways-of-acting of an object which exist as a property of that object (Bhaskar 1979). Matches light when struck across abrasive materials because they have certain properties. This understanding leads to an ontological distinction between causal laws and patterns of events; whether causal powers are evident in a pattern of events depends on contingently related factors. So, for instance, a realist might say that striking a match causes a match to light but whether the match lights depends on a number of contingently related factors, such as whether the match is dry. It is thus contingent that striking a match causes that match to light but under certain conditions it will do so necessarily. This rationale, as realists point out, underlies the logic of the experimental method, of controlling contingent factors so as to observe the causal process. For social theory, where applications of the experimental method are more limited, the implication is that causal laws refer to tendencies that may or may not be evident in events (Foot, Rigby & Webber, this volume).

This understanding leads the realist, and consequently the historical materialist, to reject two routes to theory development. First, simple enumerative induction, generalization on the basis of observed regularities, is rejected because theories are not merely generalizations but statements (abstractions) about causal mechanisms.[4] As well as rejecting enumerative induction, realists abandon the falsificationist model associated with Popper, arguing that, if causal laws are *tendencies* and not absolute predictions about events in the world, then the non-evidence of these causal processes in events is irrelevant to an assessment of the truthfulness of the posited causal mechanisms. It is partially on these grounds, the rejection of empirical regularity *as* theory (with theory here referring to the identification of necessary, or causal, connections between phenomena) and the rejection of the falsification of theory in the *absence* of empirical regularities, that the logical compatibility of statistical inference with a realist philosophy has been questioned.

I would like to argue, however, that the use of statistical inference *is* compatible with a realist view of theory. The logic of using statistical hypotheses is to generalize about empirical regularities across classes on the basis of samples of these classes. However, although empirical regularities are what statistical hypotheses are *about*, the use of statistical hypotheses does not necessarily imply an adoption of enumerative induction, as Sayer has suggested in his earlier work. An inference from a sample to a defined population or an inference that an association is such that it is unlikely to have occurred by chance does not imply a similar mode of linkage between the empirical regularity and a theory. It is possible to say that our confidence in theories is not amenable to any probabilistic treatment or wholly dependent on empirical regularity, while maintaining the importance of examining empirical regularities for theory development.

The distinctiveness between and yet interdependence of empirical regularities and theories is widely recognized within the philosophy of science. This recognition has spawned such distinctions as those between inductive evidence and inductive support, statistical evidence and statistical relevance, the latter terms in each couplet expressing a resolution distanced from the enumerative inductive mode of inference (Cohen 1970, Giere 1980, Kyburg 1970, Salmon 1975). To simplify considerably, Salmon, for instance, presents two steps in the explanatory process. The first involves statistical systemization in which empirical regularities (including those with low probability) are established. These then *demand* explanation; events and statistical regularities are fitted into a causal network which is largely an abstraction. Salmon's work follows from the understanding that the imposition of 'the high probability requirement' upon explanation – the assumption that only events occurring together with a high frequency can be explained – produces a serious malady that Kyburg has called 'conjunctivitis'. This has led to a reconceptualization of empirical regularities that exclude the high probability requirement, such that one can consider the statistical relevance of improbable events, as well as probable ones, and to an integration of this revised conception of empirical regularity with a realist notion of causality involving abstracted explanation.

It is a positive notion of support that is notably lacking in recent Marxist accounts of realist philosophy. An intriguing a priori assumption has been that associations implied by Marxist theories will *not* be evident in events. For instance, Bhaskar writes that 'laws cannot be constant conjunctions of events, because such conjunctions (1) are extremely rare, and (2) must in general be artificially produced' (Bhaskar 1979, p.162). One wonders how he *knows* this. Rather than battening down the theoretical hatches by means of a realist epistemology, it may be more productive, while embracing a realist perspective, to examine the fit between Marxist theoretical statements and empirical reality and *compare* instances of relative fit in order to understand more fully the *conditions* on which the posited causal mechanisms are *contingent*. An explicitly comparative methodology is thus advocated. Such comparisons not only indicate the generality of the associations posited theoretically, but suggest the specific contingent conditions that have led to the absence of conjunctions in particular instances.

As an example of this approach, in my own work on the links between housing tenure and political ideology in urban Canada (Pratt 1984) it was theoretically productive to compare these associations across different occupational classes. Marxist theorists, among others, have argued that home-ownership generally fosters an ethic of individualism and privatism, and that home-owners, as compared to renters, tend to centre more of their interests around the home and fewer toward issues in the workplace. An examination of the associations between home-ownership and political values across a sample representative of the adult urban Canadian population indicated that certain home-owners are indeed more conservative and less supportive of social welfare provisions, but that this generalization is limited to particular occupational groups. Specifically, a consistent pattern of associations between political values and housing tenure was found for household heads who worked in low-level managerial and skilled non-manual occupations, whereas virtually no such associations were found for those in skilled manual occupations. This finding was extremely suggestive and the reasons for this contrast between skilled manual and skilled non-manual workers were taken up in a series of in-depth interviews with home-owners and renters living in Surrey, an outer suburb of Vancouver, British Columbia. These in-depth interviews indicated that home-owner-ship did not have a critical influence on the political ideology of skilled manual workers because of their strong commitment to their trade unions. The trade unions appeared to be the critical institutional medium through which their political values were formed. In contrast, the skilled non-manual workers did not belong to unions or did not feel a commitment to those to which they did belong. In the absence of an association with a strong work-based organization, housing issues seemed to play a more important role in their political outlook. An association between political ideology and home-ownership thus seemed to be *conditional* on an absence of alternative sources of organization. As Agnew has written: 'The meanings attached to the tenure (of home ownership) differ from context to context depending on the extent to which the dominance of capitalist social relations has been successfully challenged' (Agnew 1981, p.88). The epistemological and methodological points that follow from this example are, first, that meaningful empirical regularities do exist, and second, that the comparative method is useful, both to document the generality of the empirical regularity and to uncover the contingent conditions which lead to its evidence.

This methodological prescription seems altogether consistent with Kosik's description of the historical materialist method, that is: first 'appropriating the material, in detail, mastering it to the last historically accessible detail', and then 'analysing its different forms of development' and, finally, 'tracing out their internal connections' (Kosik 1976, p.15) The beginning of the investigation is, in Kosik's words, 'random and arbitrary' (1976, p.16). Establishing empirical regularities by means of statistical inference can be seen as one aspect of this first step. The second step involves the comparative method. The final step is an act of abstraction which 'does not amount to reducing "the phenomenon" to dynamized substance, i.e. to something concealed behind the phenomena as something independent of them; rather it is cognition of the laws of movement of the thing itself' (1976, p.13). In other words, theoretical abstractions are not removed from the empirical world; they highlight the causal connections within that material world. Granted that theoretical abstractions are not empirical regularities, that they are not always evident as empirical regularities, associations at the level of events give clues to theoretical causal mechanisms and their absence raises very serious questions concerning contingent conditions. Statistical techniques are useful in assessing these patterns of association.

ATOMISM

A second and related criticism of the use of statistical inference, developed by Sayer from a historical materialist perspective, is that the ontological assumption of atomistic monism, or a variant of external relations, underlies statistical inference (although, as noted above, Sayer now argues that there is a resonance rather than a logical link between atomism and statistical inference). My intention in this section is not to argue for or against this ontological assumption. Rather, questions of ontology can be side-stepped in so far as it is possible to demonstrate that the use of statistical inference is compatible with the alternative ontology, a variant of internal relations, that is posited by Sayer. In other words, it is possible to argue that statistical inference is compatible with a broader range of ontologies than simply atomism.

Monistic atomism assumes that the world is composed of an aggregate of distinct things. Given an atomistic ontology, causation is conceptualized as a relationship between separate events, as *external relations* between independent objects. In contrast, the realist conception of causation, that of a necessary connection between events, assumes an *internal relation* between events. As an example of this type of internal relation, Sayer writes:

> It might. . . seem possible that a capitalist could stop purchasing labour-power, stop accumulating capital and therefore break our alleged necessary relationship between these actions and being a capitalist, but in acting this way s/he would be becoming a non-capitalist. In Marxism, these necessary relation-ships are 'taken up' into the definition of capital. . . (Sayer 1979a, p.16).

That is, the propensity for a certain type of action is a *property* of the capitalist and the relation between capitalist and wage earner an internal relation: the capitalist exists only in relation to the wage labourer in so far as the extraction of surplus value defines both.

Sayer is fully aware of the standard criticism of internal relations, namely, that if all relations are internally related, it is difficult to see how theoretical priority can be given to some relations over others. As a solution to this dilemma, Sayer argues that the world is comprised of a combination of internal and external relations, so that, for instance, there is a necessary relation between capital and wage labour but an external, contingent relation between capital and the raw materials used to create capital. Furthermore, conceptual necessity reflects empirically discovered natural necessity. 'When we discover such natural necessities', writes Sayer, 'we frequently make what we previously understood as contingently related elements part of the definition of objects; indeed, one might say that progress in science, in terms of reduction of the burden of facts, depends on this' (1979a, p.15). On Sayer's own account, then, we approach the world assuming that events are externally related and develop an understanding of necessarily related elements, which are so in fact, through the empirical discovery of regularities and conceptual development. That external relations are presupposed by statistical inference, used to assess empirical regularities in the course of theory development, is, then, scarcely a critique at all.

In evaluating Sayer's specific criticisms of the use of statistical inference given a historical materialist perspective, I have argued that Sayer has tended to conflate statistical inference and inductive inference (Hacking 1965), and thus the charge that statistical inference presupposes regularity generalization is unwarranted. Furthermore, the charge of atomism is incompatible with Sayer's own account of the method of discovering and integrating conceptions of natural necessity into scientific discourse.

Sayer nevertheless draws attention to the fact that specific assumptions underlie statistical testing. Along these lines, a concern with the assumption of normality which attends parametric statistics has led some 'radical' quantifiers to advocate the use of non-parametric statistics, which do not make the same assumptions concerning population distributions (Atkins & Jarrett 1979, Griffiths *et al.* 1979).[5] This suggestion seems particularly relevant given repeated claims that many Marxist concepts, including that of class, are differentiated along qualitative and not quantitative lines, although this has scarcely been considered by Marxists who have chosen to operationalize the class concept (Ornstein *et al.* 1980, Wright & Perrone 1977).

Nevertheless, some assumptions remain using even non-parametric statistics; all statistical inference assumes some notion of typicality in circumscribed populations. But surely this requirement reflects on the need for more careful theorizing about the categories compared rather than on the statistical techniques themselves. This leads to a third possible criticism of statistical inference, relevant to both historical materialist and humanist perspectives, namely that quantification and the use of statistics tend to direct attention away from the evaluation of the categories employed (Sayer 1978, Young 1979).

IMPOVERISHED THEORETICAL CATEGORIES

To continue with Sayer's earlier critique of statistical inference, he criticizes traditional industrial location analysis for 'aggregating up a large number of very diverse firms in the hope of finding "generalizable regularities" or simply to meet the requirement of an adequate sample size', because this 'has the effect of losing everything in a statistical soup. . . Apparently, making the basic categories less "chaotic" would be unacceptable since it would make the sample less "representative"!' (Sayer 1979b, p.53). In contrast to this approach, Sayer notes the analytic clarity of the work of Massey & Meegan, suggesting that it is due, in part, to their 'avoidance of empiricist generalization'.

Sayer may present an accurate assessment of the way in which statistical techniques have tended to be used in locational analyses, but it is an overstatement to suggest that statistical inference necessarily entails a 'chaotic' aggregation of categories. Statistical techniques are used to measure patterns of association between categories and populations that an investigator defines. The fact that traditional locational analysts have tended to aggregate diverse firms reflects on the inadequacy of their theoretical development rather than on any weakness of statistical techniques *per se*. There is no logical reason why a sample of firms must be representative of all firms, rather than, for example, firms that are restructuring to achieve scale advantages.[6] Once again, statistical inference is analytically distinct from inductive inference. In the former instance one generalizes *only* to a theoretically defined population. It is the task of the analyst to define the population in a theoretically meaningful manner.

Although shifting the focus from statistical inference to the use of descriptive statistics, it is none the less of interest that Abrams has recently argued that the principal value of quantification in history has been as a catalyst for a radical questioning of received categories. Referring to Laslett's discovery of the pre-industrial 'reality' of the nuclear family using quantitative techniques, he writes:

> [Laslett's] achievement is to have renewed the history of the family as a problem. . . . But what it contributes to our knowledge of how it was is not so much new facts as new doubts, a more complex sense of the relational realities standing within the social forms directly available to empirical scrutiny. The problem of how it was runs beyond the observational techniques of quantitative data gathering into a distinctly theoretical terrain (Abrams 1982, pp.321–2).

Abrams further outlines the necessity of going beyond the quantitative base in relation to McFarlane's inability to discover a peasant society in pre-modern English history. In relying on a head count of peasants as a *measure* of feudal society, Abrams argues that McFarlane fails to make important analytic and relational distinctions. McFarlane gives too much *theoretical* weight to the observed frequency statistics and fails to consider the extent to which the relation between lordship and peasant can nonetheless be shown to have been the operational pivot of pre-modern English society (Abrams 1982, p.325). Nevertheless, the quantitative evidence presented by McFarlane leads to a refinement and sharpening of the concept 'peasant' which, Abrams (p.326) judges, had previously been used 'so inclusively as to deprive it of much of its possible analytical value'. Again, the important interdependence between quantified empirical reality and theoretical development is noted, with the proviso that one is not synonymous with the other.

If Abrams stresses the influence of quantified data on theoretical categories, the other half of the interdependent relationship has been equally stressed. At base, numbers are always measurements of theoretical categories which, in turn, cannot be separated from their social context.[7] This returns to the point that meaningful quantified analysis, even the use of descriptive statistics, is dependent on adequate theoretical development. The point that numbers are produced, not simply collected, has been made forcibly by a variety of authors, especially in relation to official statistics (Armstrong & Armstrong 1983, Atkins & Jarrett 1979). Armstrong & Armstrong, for instance, demonstrate the way in which the Canadian census systematically distorts the employment situation of women through the questions excluded, the wording of the questions included and the categories designated to code the answers (Armstrong & Armstrong 1983). For example, the definition of work, which excludes housework and volunteer work but includes unpaid employment in a family business, reflects the assumption that work which results directly in profit and exchange is of unique concern. The census data on *paid* female employment must also be interpreted in the light of the social context. In-depth interviews suggest that certain institutional constraints in Canada, such as low minimum earnings set for daycare subsidies, encourage under-reporting of employment and earnings by low-income women. These comments suggest that quantitative data must always be interpreted in the light of the theoretical and societal context in which they are collected. For this reason quantitative data are usefully complemented by more qualitative understanding of the context.

At issue here is not only depth of understanding but the sheer accuracy of interpretations based on quantified data. Ley cites Brian Berry's interpretation of the increasing correspondence between the city-size distribution of Israel and the rank-size role through time as one example of the erroneous deductions one can make from descriptive statistics which have been removed from their context (Ley 1978). Berry interprets the noted historical patterns as evidence that increasing economic complexity leads to a closer approximation to the rank-size rule. Ley points out that the national boundaries of Israel were radically altered during the time period under consideration and that the increasing approximation of the city-size distribution to the rank-size rule reflects the redistribution of cities across political boundaries rather than increased economic complexity. Only a contextual understanding of the Israeli situation leads to an accurate reading of the descriptive statistics.

An understanding that quantitative techniques, including statistical inference, are but one type of evidence and that theoretical conceptions infuse empirical evidence, while neither is reducible to the other, is especially important given the final criticism of quantification to be considered, that it necessarily leads to a neglect of meaning, human intentionality and choice (Jarrie 1983).

REDUCTION

For the present purposes, a number of complex and somewhat distinct arguments concerning the antithesis between quantification and a non-reductionist conception of human nature will be combined. On one level, it is argued that the precision required by the quantification process serves to restrict meaning to discrete categories and forces a unidimensionality that obfuscates an understanding of relational processes. A related point is that the notion of an active and integrated social agent is lost by a perspective that isolates, and then perhaps reifies, particular quantifiable characteristics (Ley & Samuels 1978). Further, in line with the criticism of an overriding obsession with empirical regularity, an exclusive concern with the general over the unique runs against the importance of human agency, however constrained, in effecting social change (Samuels 1979).

These criticisms point to the profound limitations of measurement rather than an outright rejection of quantitative techniques. For instance, empirical regularities may point to patterns of social activity that can then only be understood by grasping the meaning of these actions in a particular social context. The weight of the humanist critique is that some properties are not measurable in principle; this does not imply that no properties are measurable or that such measurement is irrelevant to our understanding of social processes.[8] What the humanist critique states is that relationships between measured quantities are one type of evidence, which must be understood and conceptualized, in the situational context.

Arguing for an existential geography, Marwyn Samuels has noted that it is:

broad enough to encompass not only the 'geosophy' of a J.K. Wright or the 'topophilia' of an Yi Fu Tuan but also the geometric modeling and distance measures of a Walter Izard, a Brian Berry, or a Peter Gould. . .Where the one is an exercise in rootedness, the other is an exercise in rootlessness. They are not opposites but correlates; and, in existential terms, the one (a finding of roots) is but the outcome of the other (a loss of roots) (Samuels 1978, pp.35–6).

Giddens similarly points to the complementarity of quantitative and more qualitative understanding:

So however statistical a given study might be, however abstract and remote it is, it presumes some kind of ethnography of individuals involved in the context which is being described. That ethnography is not, of course, always overtly written: in a statistical study, for example, it's largely a tacit taken-for-granted aspect of what's done, but it is necessarily there, I think, and it could in principle be written (interviewed by Gregory 1984, p.129).

Both the humanist and historical materialist perspectives suggest that we must seriously respect the distinction between quantitative and more qualitative analysis, and between empirical regularities and theoretical conceptions, not in an absolutist, either-or fashion but in the sense of realizing the limitations and restricted appropriateness of particular methods and types of knowledge. As Berka has recently noted:

Mathematization in science has only an instrumental character. The usefulness of this approach depends on its thoughtful usage. . .as well as on its appropriate specification with regard to the concrete object of investigation. . .it is always only a means of helping to objectivize the results of inquiries in the social scientific domain in question (Berka 1983, pp.216–17).

The helpfulness is this process of objectification must not be lost through a confusion of statistical inference and quantification with the broader philosophical perspective of positivism.

CONCLUSION

It has been argued that statistical inference and, more generally, quantification, are not susceptible to several of the criticisms that have been directed toward them by historical materialist and humanistic geographers. That is, the use of statistical inference does not necessarily imply the reduction of theory to empirical regularities or the ontological assumption of atomistic monism. It need not necessarily lead to impoverished theoretical categories or a reductive, mechanistic view of human nature.

At the same time, the value of the criticisms outlined by historical materialist and humanistic geographers is that they emphasize the distance between quantitative research appropriate to these perspectives and that conducted within a positivist philosophical framework. They draw attention to the fact that statistical analysis cannot subsume theoretical development, that theoretically uninformed quantitative research is of little interest and of little use in explaining social processes. In other words, such criticisms serve to pare quantitative research down to size, to the status of a technique for organizing one type of information, information that needs necessarily to be complemented by a more relational, contextual understanding, as well as more abstract theoretical development.

NOTES

This is a revised version of the paper, 'Does philosophy matter? Humanistic-materialist perspectives and quantitative techniques', presented at the annual meetings of the Canadian Association of Geographers at Nanaimo, British Columbia in May 1984. I would like to thank Susan Hanson and Suzanne MacKenzie for helpful comments on the initial draft.

1 Sayer has modified this position is his more recent work in which he suggests that there are 'resonances' rather than logical necessities which encourage a clustering of certain philosophical positions and techniques (Sayer 1984).
2 The domain of this chapter does not include an evaluation of these criticisms. For useful summaries see Gregory 1978, Keat & Urry 1975.

I leave untouched, for instance, the issue of whether positivists accept the weight of these criticisms.

3 On very different grounds, Couclelis & Golledge (1983) also argue that analytic geography shares some common themes with both the phenomenological and Marxist perspectives.

4 It is perhaps worth noting that modern realists are not distinct in this assessment; in the 17th century Francis Bacon wrote that 'induction that proceeds by simple enumeration is puerile' and advocated the construction of abstract models of the world to go beyond sense data or, rather, a variant of abduction (Bacon, quoted in Hacking 1975, p. 16). It is irrelevant to *this* argument whether positivists would agree with this characteristic of their theory development. The realist position is being developed here to show that statistical inference is not incompatible with the realist view of theory.

5 This type of caution has come from within the positivist tradition as well. See Gould 1970.

6 The latter analytic category is one used by Massey & Meegan 1978.

7 As Susan Hanson (personal communication) pointed out, this issue is not new but, rather, underlies the whole question of validity in measurement.

8 This view is consistent with a historical materialist approach to measurement in so far as emphasis is placed on measurability as a *real* property of phenomenon and processes (Berka 1983). This point is particularly well developed by Sayer 1984.

6

Theory and measurement in historical materialism

SIMON FOOT, DAVID RIGBY
& MICHAEL WEBBER

INTRODUCTION

There are three kinds of activity in a scientific research programme: elaboration and modification of a body of theory, use of that theory to explain actual historical events, and production of a philosophical basis for the procedures in the research programme. It has recently been argued that Marxists adopt realism as a philosophical basis for their research programme (Chouinard *et al.* 1984, Williams 1981), a claim that implies particular links between the use of the theory in explanation and its modification after evaluation. This chapter tries to state and to illustrate the way in which Marxist theory is evaluated by reference to experience, that is, actual historical events.

The question of the scientific nature of Marxist geography has been addressed by several recent papers. Sayer (1982a) has described a realist basis for science and has attacked the kind of science performed by non–Marxist geographers. By contrast, Couclelis & Golledge (1983) question the description of Marxist geography as science because for them it lacks formal evaluation procedures for its theory. Whereas Papageorgiou (1982) enters a plea for tolerance in this dispute, Chouinard *et al.* (1984) have argued that, in practice, Marxist and non-Marxist geographers use very similar evaluation procedures notwithstanding their different 'official' philosophical positions. Our chapter is concerned with a very small part of this debate, namely to show how abstract laws and concrete events may be linked to provide an evaluation of theory.

There are many different aspects of a theory that must be evaluated. First, its logic must be analysed for internal consistency. Second, the theory must be shown to be consistent with existing, non-competing theories and existing knowledge. Third, the theory must not be derived in an *ad hoc* fashion from existing theories. Marxists also apply a political evaluation of theory: can it be used in the long or short term to enhance the power and position of the underprivileged? In this chapter, however, the issue is empirical evaluation. In particular we have two aims: to show that it is possible in some circumstances to measure the contribution of different mechanisms to an outcome and to show how our knowledge of these mechanisms can be tested by evaluating their use in explaining actual historical circumstances.

The essential problem is this – and it is the same for all non-experimental scientists. Theory claims that in particular circumstances an event (x, say) should occur. In fact, during one period of observation, when circumstances are different from those postulated in the theory, y occurs. Can that theory be held then to help explain y, the actual history? The second section lays out the argument that, with appropriate measurements, an attempt can be made to use the history to evaluate the theory even though actual circumstances differ from those postulated. The third section illustrates this argument using the example of the development of the current economic crisis in Canada, and the fourth section employs the example of the postwar history of the steel industry in North America to illustrate how to pass from this abstract analysis to examine more concrete events.

METHODS OF EVALUATION

Realist philosophy makes several claims that distinguish realist from non-realist methods of scientific enquiry (Bhaskar 1975, 1979). The three central claims are that a structured world exists independent of human thought, that causal mechanisms generate empirically observable outcomes in that world and that these mechanisms are ontologically distinct from the outcomes that they produce (Bhaskar 1975, p.15, Engels 1934, p.218). Furthermore, scientific laws or causal mechanisms (statements about logical necessity) are not empirical regularities but are tendencies. Even though a causal mechanism is operating, the type of event to which it gives rise is influenced by other causal mechanisms, and also depends on the specific conditions that pertain at different times and in different places. There is therefore no necessary one-to-one correspondence between empirical outcomes and a particular causal mechanism: this is what it means to say that mechanisms and their observable outcomes are distinct.

One of the aims of science therefore is to identify causal mechanisms despite the variety of forms in which they reveal themselves. This is made more difficult in social science than in most physical sciences because of the impossibility of constructing experimental conditions. It is not possible in social systems to isolate, as in a laboratory, the historical and locational conditions within which a causal mechanism can be seen to operate. Identifying causal mechanisms in social science is also more difficult because of the intervention of human agency which, among other things, can change causal mechanisms. For the purposes of this chapter, however, it is taken as given that the causal mechanisms central to societies that are organized around a capitalist mode of production are class struggle and inter-capitalist competition.

Our aim here is to show how these causal mechanisms, and the variety of concrete events which are determined by them, can be linked to provide an evaluation of theory. To fulfil this task we employ the example of the theory of the rate of profit, and in the following sections describe its influence on the postwar history of profit rates in Canada and the decline of steel production in the United States. (For a review of arguments about the falling rate of profit, see Shaikh 1978, Wright 1979; and for evidence about the history of the rate of profit in Canada, see Webber & Rigby 1986.) In order to do this it is first necessary to specify exactly what we mean by a tendency. In other words, why is it that there is no one-to-one correspondence between class struggle and capitalist competition – the causal mechanisms – and the variety of events to which they help give rise: why is it that a given necessary relationship, or law, does not produce a regular outcome? The literature does not provide a consistent answer to these questions.

Bhaskar offers two philosophical reasons for the lack of correspondence between a given law and the variety of its outcomes. First, there exist several different causal mechanisms. In physical science these forces are transcendental: they always hold, though some may be more influential than others in the causing of particular events. Therefore it is not possible to credit to all events the same combination of mechanisms as their causes. Second, the arrangement of things that make up the material world is such that a variety of different concrete conditions can arise in which laws combine to produce a series of irregular events. In other words, outcomes are not necessarily determined in a regular fashion by laws, but also depend on the way in which the operation of those laws is combined in a particular conjuncture.

Take as an example of this view of tendencies the causal mechanism of gravity, which we shall define as the relationship of attraction between objects. (An example from the physical sciences is easier to think about because we can imagine its operation within a defined set of conditions, or in a closure.) The law of gravity states that objects attract each other with a given force that depends on their mass and distance apart. But this does not always give rise to the same event. First, the conditions that pertain to different places and times vary. So a baseball which is dropped on earth will behave in a way quite different from one dropped from the same height on the moon. This is because the conditions surrounding the ball – the masses that attract it – are quite different in each place, though the law of gravity operates in just the same manner in both places. But on the earth another causal mechanism also intervenes, because there is friction between the air and the ball which slows its progress towards the earth. Thus the law operates in different conditions and gives rise to a variety of events.

In social science, however, the concept of laws as tendencies has not been treated so strictly. This is partly due to a misuse of the concept, but it is also due to a fundamental distinction between the method of science proposed by Bhaskar and that inherent in the work of Marx and many neo-Marxists.

In the first place, the concept of a tendency has been misused to describe an historical trend. Jessop, for example, refers to tendencies directly as historical trends, such as those 'towards the socialisation of labour, the concentration and centralisation of production (and) the relative impoverishment of the working class' (1982, p.72). These 'tendencies' are thus given an inherently empirical content by Jessop, whereas the term is meant to convey the distinction between abstract theoretical relationships and the *variety*, not the similarity, of the empirical outcomes they influence. An identified 'pattern' in history (e.g. people are moving toward the suburbs) alone is not a tendency but a trend: suburbanization is an outcome of causal mechanisms and contingencies, not a law about capitalist development expressed as a tendency. Thus we should speak of the trend toward suburbanization rather than the tendency toward suburbanization; of the trend toward the growth of state power; of the trend toward de-skilling of the labour force. Equally, throughout the postwar period the rate of profit in Canadian manufacturing industry exhibited a marked downward trend.

The second use of 'tendency' in the Marxist literature provides a significant criticism of Bhaskar's conception of causality. This is the characterization of tendencies as forces that lead to contradictory outcomes, a view which acknowledges the dialectical nature of history. Such a view implies that actual history is not simply the outcome of unidirectional or mechanistic causes, but results from conflicts between opposites. Furthermore, causes do not merely exist at an abstract level, but are embodied within the concrete events which are their outcomes. It follows that within each concrete stage of history is embodied the conditions that give rise to the forms of subsequent outcomes.

This argument leads us to modify Bhaskar's account of realist science. It suggests that the conditions in which causal mechanisms operate at any one time are not themselves a matter of chance but are both the outcomes of previous stages, or specific periods of development, as well as the causes of subsequent events, in so far as they influence the degree to which each mechanism operates. History is therefore evolutionary, not a series of unconnected, specific events. This also implies that in the production of each conjunction of events exist the conditions for their own destruction.

We are now forced to accept that events are not simply the result of invoked causal mechanisms in specific historical and locational situations. They are also determined by the course of history, so that particular conjunctions of events are implicitly both the outcomes of previous developments and the causes of those to follow. Therefore, any interpretation of the various forms of outcomes must not only refer to their abstract causes and historical specificity, but must also refer to their place in the course of history. So, when we refer to realist laws we imply not only that the forms of outcomes or events are different because causal mechanisms work in different combinations in different situations, but also that the events themselves alter the conditions which prevail, thus providing for the emergence of new outcomes.

There are two more things to say about tendencies. First, our discussion does not distinguish between tendencies and counter-tendencies, the latter a term used in the literature to refer to some group of events or actions which appears to oppose the effect of the tendency. Marx's discussion of this issue, for example with respect to the tendency for the rate of profit to fall, is confused; he includes in the list of counter-tendencies both theorized effects, such as a reduction in the value of commodities, and trends, such as increases in foreign trade. But in making this distinction we note that forces which operate to raise the rate of profit are the effect of causal mechanisms just as much as the forces that reduce the rate of profit. The outcomes of the two tendencies then work in different directions and are concentrated at different times. It is not, however, clear that one force is necessarily secondary to the other, as implied by the tendency/counter-tendency terminology.

In order to clarify these arguments we examine the theory of the falling rate of profit. In a two-department model of the capitalist mode of production, the rate of profit for a given production period is defined as

$$\pi = \frac{\text{SURPLUS}}{\text{CAPITAL ADVANCED}}$$

$$= \frac{t(1 - \lambda_2 D)}{\lambda_1 K/L + \lambda_2 D} \qquad (1)$$

where

π = rate of profit per year
λ_1 = unit value of the means of production
λ_2 = unit value of the means of subsistence
D = real wage per unit of labour
K = amount of means of production used per production period
L = amount of labour used per production period
t = number of turnover periods per year.

Thus,

$\lambda_2 D = \lambda_L$ = unit value of labour power
$\lambda_2 DL = \lambda_L L$ = value of variable capital
$1 - \lambda_2 D$ = surplus value per unit of labour power
$\lambda_1 K$ = value of constant capital
$\dfrac{K}{L}$ = technical composition of capital

The magnitudes of the terms in equation (1) determine the rate of profit. The problem for theorists, however, is to explain how these terms will change in response to conditions in society. To do this, we must explain how the central causal mechanisms of capitalism (class struggle and inter-capitalist competition) produce changes in the variables and so in the rate of profit.

Class struggle is manifest in many ways and its ramifications extend throughout society. We may identify, for example, attempts by labour to reduce the degree to which labour power is a commodity, to raise the wage rate and to improve the quality of life, and efforts by capitalists to maintain a commodity economy and to reduce the value of labour power. There are of course many other issues of struggle, but their combined significance to the determination of the variables in equation (1) is that they affect the rate and direction of change of real wages, that is to say, capitalists and labourers struggle over the magnitude of the real wage.

The second mechanism of capitalist society is inter-capitalist competition. Labour produces profit, but capitalists compete with each other in an attempt to obtain as large a share of the surplus produced as possible. There is a variety of ways of doing this. In production, capitalists compete by minimizing costs, by improving productivity and introducing more efficient techniques of pro-duction. Some firms compete on the strength of product quality. In the market capitalists compete by attempting to purchase inputs at prices below values and by attempting to sell outputs at prices above the market average.

Class struggle and competition do not operate in isolation. However, these mechanisms may vary in relative strength, depending on circumstances, so the forms of outcomes which they generate vary. There is no one-to-one correspondence between the mechanism and the events, even to the extent that collusion between capitalists and alliance between classes are plausible outcomes. These arguments are briefly illustrated by two examples.

First, suppose that the rate of profit is high. Then the rate of accumulation can be high in relation to the rate of growth of the labour force. This implies that the balance of class strengths swings towards labour and therefore that the real wage tends to rise. It also implies that capitalists have therefore an increased incentive to replace labour with machinery as a way of reducing costs. Consequently, in these circumstances, the mechanisms of class struggle and inter-capitalist competition produce twin effects: the struggle over the real wage is manifest as a trend toward rising real wages while attempts to reduce costs appear as a rising technical composition of capital. As is evident from equation (1), these manifestations of the laws cause the rate of profit to fall.

Suppose, second, that the rate of profit is low. In this case, the rate of accumulation must also be low and some firms will go bankrupt. Therefore labour becomes relatively more abundant in comparison with the accumulation of capital and the balance of class strengths swings against labour. The struggle over the real wage is revealed now as reductions in real wages. Profits being low, capitalists have a strong incentive to reduce costs but because wages are relatively low, they have little incentive to replace labour with machinery. In this circumstance, the laws are manifest in changes that cause the rate of profit to rise.

Thus we see that outcomes are produced by the operation of a mechanism in a particular context. There is always struggle over the real wage and capitalists are always seeking to reduce costs. But the forms this takes depend on circumstances; this is what it means to say that there is no one-to-one correspondence between a mechanism and an observable outcome. Class struggle is manifest as a rising or as a falling real wage, depending on the relative rates of growth of capital and of the labour force. Equally, reductions in the costs of production may or may not be associated with a rising technical composition of capital, depending on the relative prices of capital and labour.

This example provides several lessons. First, causal mechanisms provide the structures within which individuals make decisions and their history. These decisions are not predetermined by the mechanisms but also depend on a host of contingent factors. Such factors may include the personal psyche of the individual agents, how they perceive their environment and also how the constraining structures are randomly modified by 'accidental' features of space and time.

Second, the causal mechanisms and actions they prompt cannot be considered ahistorically. Individual responses to forces operating at a particular juncture influence how these forces reveal themselves at another moment in time. For example, although new forms of labour management introduced in the early 20th century, such as Taylorism and Fordism, were immediately beneficial to capitalist accumulation, they ultimately led to new forms of conflict between capital and labour, necessitating a quite different response in the form of bureaucratic control and collective bargaining (Burawoy 1979, Edwards 1979).

Realist Marxists begin with this philosophy of causes, construct a theory that identifies a set of causal mechanisms and the events that they produce in particular contexts, and try to evaluate that theory. The empirical part of this evaluation is accomplished by testing the ability of postulated causal mechanisms to explain particular events as the combination of laws, contingencies and historical contexts. Like Lakatos (1978), we require that the continual modification and elaboration of the theory that is required by the partial success of these evaluations shall be creative rather than *ad hoc*.

The following two sections illustrate the arguments above. The third section documents the postwar history of Canadian manufacturing performance in terms of abstract forces and relations. The development of a crisis in the Canadian economy is linked to a reduction in the general rate of profit. The fourth section examines one part of this crisis, its manifestation in the North American steel industry, and traces out the forms of mechanisms and contingencies as events that must be assembled to explain the recent history of that industry.

THE RATE OF PROFIT IN CANADIAN MANUFACTURING, 1950–81

In this section Marxist theory and empirical investigation are combined to explain movements in the rate of profit in the Canadian manufacturing sector between 1950 and 1981. The first part of this task involves measuring the components of equation (1). The second problem is to interpret the information gathered. Our interpretation of the data is based on Marxist economic theory. In particular we show how that theory can be used in a logical manner to explain the fall in the rate of profit and the economy's slide toward a recession or crisis.

Marxist economic theory argues that economic growth cannot continue to increase at an uninterrupted rate. The basic claim is that as the economy expands faster than the rate of growth of the labour force, workers find themselves in an increasingly advantageous position and can demand higher wages. Capitalists are then forced to adopt labour-saving technical changes in order to reduce the demand for labour. Innovations of this type, however, cause the technical composition of capital to rise and, in the absence of offsetting factors, the rate of profit must fall. Though individual firms and even industries can withstand such pressures if they are more efficient than average, the economy as a whole is

Table 6.1 Manufacturing rate of profit

Year	D	K/L	λ_1	λ_2	t	$\pi(\%)$
1950	49.5	46.2	143.7	213.5	88.6	45
1955	62.3	56.8	123.8	156.6	95.5	45
1960	73.1	73.0	117.7	142.4	91.0	37
1965	80.1	73.9	119.3	116.7	103.4	37
1970	95.6	95.0	105.7	110.6	98.9	32
1975	112.8	114.9	93.8	91.5	97.7	30
1980	120.5	128.2	87.5	79.6	106.5	29
1981	119.0	130.3	89.3	79.3	108.0	29

Note: The rate of profit is given as a percentage. All other data are index numbers with a base of 100 in 1971. For sources see Webber & Rigby (1986).

forced to undergo periods of slower growth or recession (Shaikh 1978, Weisskopf 1978).

Table 6.1 documents the values of the components of the manufacturing rate of profit in Canada between 1950 and 1981. For a review of the sources and methods used see Webber & Rigby (1987). The table shows that the real wage increased strongly between 1950 and 1981. The rate of increase was most rapid in the early 1950s averaging about 5% per annum. From the mid-1950s to the mid-1970s, the real wage increased at approximately 3.25% per annum and thereafter decreased. A measure of technical change in the economy is provided by the technical composition of capital. Our theory argues that this variable should increase as real wages rise. In fact Table 6.1 reveals that the technical composition of capital increased by almost 300% between 1950 and 1981. The process of accumulation in postwar Canada led to increases in both real wages and the technical composition of capital, both of which trends exert a downward force on the rate of profit (see equation (1)).

It would be wrong however to claim that a falling rate of profit is the necessary outcome of the underlying mechanisms of class struggle and competition that structure the capitalist mode of production. This is because capitalists can adopt a variety of actions which, in spite of increases in the real wage and the technical composition of capital, may increase the rate of profit. The routes capitalists may follow are revealed in equation (1). We have seen increases in the real wage and the technical composition of capital tend to depress the rate of profit. However, if capitalists can decrease the unit value of capital and wage goods (λ_1 and λ_2 respectively) or decrease the turnover time of capital, then the rate of profit may rise. To complete our investigation of the rate of profit the changes in the values of commodities and the rate of turnover of capital in the Canadian manufacturing sector between 1950 and 1981 must be examined.

From Table 6.1 observe that the unit value of the means of production (λ_1) fell from 1950 to 1981 by 38%. It was this decline in the value of the means of production that formed the basis for the relative cheapening of both capital and consumer goods. Table 6.1 also illustrates the decrease in the unit value of the means of consumption (λ_2). Between 1950 and 1974, the value of consumer goods fell at a faster rate than the value of capital goods. Capitalists also managed to increase the rate of turnover during the period examined, the number of turnovers each year increasing from four to just under five between 1950 and 1981. The decreasing values of commodities and the rise in the rate of turnover exerted an upward pressure on the rate of profit.

The rate of profit is the synthesis of the forces identified above. Lastly, Table 6.1 shows that between 1950 and 1981 the rate of profit in the Canadian manufacturing sector fell from 45% to 28%. This indicates that increases in the real wage and the technical composition of capital outweighed the influence of reductions in commodity values and the increasing rate of turnover of capital. We argue that the abstract forces discussed earlier were responsible for the fall in manufacturing profitability, and hastened the onset of a crisis in the Canadian economy.

Several contingent factors however compounded the problems of Canadian manfacturers. First, a history of underdevelopment in the production of capital goods made the manufacturing sector weak and reliant on foreign capital. Second, a decrease in demand for Canadian products on world markets as foreign economies became more competitive meant that the slack could not be taken out of the domestic market. Third, rising fuel prices in the world market meant higher costs in the manufacturing sector.

This explanation of crises is useful at a general level. It begins with a series of logical deductions from our knowledge about the structural causal mechanisms of capitalist production, the central claim being that when the rate of profit in the economy is high we would expect it to fall. In other words there are certain kinds of general outcomes we would expect given a certain set of conditions. We then compared these expectations with empirical evidence of postwar trends in the Canadian economy. Our theoretical arguments were found to be consistent with these facts. Such investigation provides a useful test of our theory. Had the forces identified led us to believe that the rate of profit should have risen since 1950, when in fact it fell, we would have been forced to question the efficacy of our theory (assuming no empirical error).

Still, the evidence in this case leads us to believe that we have successfully identified the causal mechanisms and a general set of conditions that produced a crisis in the Canadian economy. The third section, however, only points us to the next stage of explanation. Primarily our argument so far rests on logic. In other words, we would expect the same kinds of events to occur in any capitalist economy facing similar conditions. What this argument fails to do is identify any difference between one crisis and another.

To advance our explanation further we need to examine the crisis at a more specific level. This does two things. First it shows how abstract causes are actually produced 'on the ground' by the actions of individual people, firms and political groups. Second, through historical analysis which traces the development of a series of events, it shows how these events themselves alter the specific conditions within which the causal mechanisms operate. It is in response to these alterations that the outcomes of mechanisms change in form, even though the mechanisms themselves remain the same. The abstract and the concrete are integral parts of explanation. The completion of our explanation of crisis is now illustrated with an analysis of the decline of the US steel industry since 1950.

US STEEL IN DECLINE

This section examines the development of a crisis in the US steel industry. In general, the explanation of this crisis draws on the abstract forces identified earlier. However, the analysis here also reveals that a thorough explanation of the history of the steel industry must go further and show how individual agents responding to particular pressures made that history.

In 1950 over 45% of the world's raw steel was produced in the USA and at that time US integrated steel mills were the biggest, newest and most technologically advanced in the world (Barnett & Schorsch 1983, p.23). The large domestic market had permitted American firms to scale their operations efficiently, an advantage that was unattainable by other producers so long as their markets were limited by national boundaries. The European and Japanese economies were still recovering from the destruction of the 1939–45 period, while the US industry enjoyed the opportunities for expansion provided by the Second World War and later the Korean War.

Investment in US steel production in the 1950s focused on scale increases, not technological improvement (Barnett & Schorsch 1983, p.28). But in the early 1960s two major technological developments became generally available to steel makers: the basic oxygen furnace, which cut to one-third the time required to change molten iron and scrap into steel, and continuous casting, which eliminated the time-consuming and relatively labour-intensive process of casting ingots. Thus, the US steel industry found itself in the 1960s capitalized with outdated and as yet undervalorized machinery, while the economies of Europe and Japan were now in a position to invest more heavily in new steel-making equipment.

At the same time shipping improvements began to open up cheap sources of iron ore in Australia and Brazil, as well as foreign markets, to producers in Japan. Japanese steel makers also had the advantage of cheaper labour. Barnett (1977) estimates that wages in Japan were one-fifth of those in the US in 1964. In the 1970s new steel plants had been installed in various third world countries, particularly Korea and Brazil, where wage differentials were even greater (Barnett & Schorsch estimate a tenfold difference between the US and Korea), and such countries are emerging as significant steel exporters. In 1983 Korea and Brazil were the third and fourth largest exporters of steel to the US, behind Japan and West Germany (American Iron and Steel Institute, 1983).

Whereas the US was a net exporter of steel in the 1950s, by 1981 some 25% of the US market was captured by imports (Manion 1983a, p.23). For the first quarter of 1982, shipments from the domestic integrated steel producers were down to 17 million tons from 23.4 million in the first quarter of 1981. Of the 75 million tons of American carbon steel capacity, about 50% was idle at the end of May. Some 20 million tons were in jeopardy of remaining closed and were up for sale or slated for permanent closure in the forthcoming year (*Business Week*, 1982). By the end of the first quarter of 1982 imports had reached 4.9 million tons, up 34% from the same period in 1981, so imports were actually increasing their share of a declining market. By the end of the third quarter capacity utilization had fallen to a mere 30% (Manion 1983a).

Explanations of decline in the US steel industry typically credit that decline to a misguided government policy (Manion 1983b), or to management decision errors (Barnett & Schorsch 1983). Such an approach however provides little more than a description because it lacks any consideration of determination. The government made its decisions in the context of its budget deficit, foreign policy and public opinion it is true, but to provide a complete explanation we have to understand the decline of steel production in the USA in the context of real, underlying structures. In the previous section we established that a rise in the technical composition of capital tends to reduce the rate of profit in the economy as a whole. This does not mean, however, that those firms that adopt a new technology suffer reduced rates of profit. Though the adoption of new, labour-saving techniques causes an overall fall in the rate of profit, it is in those firms which fail to adopt the new technology that the overall fall in profit rates is apparent. This is observed in the steel-producing firms of the United States where technical advance has been slow. Firms which are making losses are located in the United States rather than in Japan or Korea.

The decline in US steel production is not simply the result of intensified foreign competition. There is also evidence of struggles between the steel sector of capital in the US and its labour force which compounded a growing inability to counter a fall in profits. Throughout much of the study period, wages in the US steel industry remained high, a direct result of the actions of a strong union – at least relative to unions in Japan and third world countries – that gained substantial political power during the upward swing in capitalist accumulation in the 1950s and 1960s. Furthermore, labour in the steel industry was struggling against an increasingly weak sector of capital under steadily rising competitive pressure from abroad. The three-yearly negotiated contract with United Steel Workers of America (USWA) led periodically to labour unrest, most notably during the nation-wide strike of 1959. It was particularly during these periods of unrest that imports gained new footholds in the domestic US market (Mueller 1982), so the contract signed in 1974 was the result of a specific attempt by steel management to ensure an uninterrupted stream of production in the years to come. In this contract (the Experimental Negotiating Agreement or ENA), the right to strike was dropped, in exchange for costly wage concessions and improvements in other forms of compensation (Betheil 1978).

Of course the logical strategy in the face of a high wage environment for US steel capitalists would have been the adoption of labour-saving technology. However, it was becoming increasingly difficult for the steel corporations to attract the huge amounts of financial capital required to regain technological competitiveness. This was due to the declining profit rates in US steel production, the cost of abandoning large amounts of outdated but as yet undepreciated fixed capital, a sluggish domestic market, and competition from cheap imports, produced with cheaper costs, efficient methods and passive labour. Therefore the crisis in US steel production emerged within the context both of manifest failures to maintain control of the labour force and of superior competition from other steel makers abroad.

The primary reaction of American steel producers in this situation has been to turn to the state for assistance. But with the exception of the post-1980 period, the relationship between the steel industry and the US government has been anything but cordial. Despite lobbying by steel management for protection against supposedly underpriced or subsidized steel from other countries, tariff regulation has been intermittent and predominantly unsuccessful. The 1971 voluntary trade agreements (Government Accounting Office 1974) were aimed at steel by tonnage not price, and therefore succeeded only in shifting import penetration to high grade or speciality products. The 1977 trigger price mechanism (Government Accounting Office 1980) aimed to eliminate dumping practices by government-assisted steel industries abroad, but failed to stem the increasing flow of cheaply produced steel from low-cost production locations, in particular from developing countries. The continued lack of effective market protection represented by these measures betrays a primary concern of the US government to keep steel prices low for other domestic producers, an important consideration for the health of the economy as a whole since steel is a sector of production with large forward linkages. This attitude was also evident in vigorous attempts during the 1960s by the US government to force down the price of domestically produced steel in the light of a perceived oligopolistic pricing mechanism in the industry. In the US, depreciation allowances for tax purposes were between 15 and 18 years until 1981, in comparison with 1 and 2 years in the UK and Canada respectively (Barnett & Schorsch 1983). The characteristics of a developed capitalist economy in the 1950s and 1960s, including the density of industry and a high standard of life, can be credited with forcing, through the state, the introduction of limited environmental controls. But limited as they may be, their introduction required $4 billion of expenditure by the end of 1979 and according to Barnett & Schorsch (1983) a prospective further $4 billion in the following four years. Crandall (1981) has estimated that some 6% of the cost of US steel in 1983 may have been due to environmental expenditures. At least for the steel industry these have been entirely unproductive investments.

So state policy, reflecting the interests of labour and, more particularly, the interests of capitalists in other sectors of production, has not been favourable to a steel industry recovery. Profit rates have fallen consistently as wage rates have remained high, as the US cost advantage in raw materials has been eroded through shipping improvements, and as the US industry in general has fallen to the bottom rung on the ladder of technological efficiency. The introduction of new technology in other countries places pressure on the general rate of profit, but it is in those firms failing to make such changes, a characteristic of many US steel firms, that the rate of profit actually falls and closure is common.

CONCLUSION

The methods of evaluating theories which are based on a realist philosophy of science are outlined and illustrated in this chapter. They are different from those used for many non-realist theories because realism does not claim a direct correspondence between causes and effects. Thus, we argue in general that it is not possible to test a theory by its ability to predict concrete events.

Our arguments in the second section are built upon the work of others who have established rules for realist analysis (Bhaskar 1975, 1979, Chouinard *et al.* 1984, Keat & Urry 1975, Sayer 1982a). But we make an original contribution in clarifying what is meant by the term 'tendency'. It describes the relationship between causes outlined above, and as such is an epistemological term. When used in a substantive context, which it commonly is in the literature, as for example in the 'tendency for the rate of profit to fall', it leads to confusion. This is because it then suggests a relationship between cause and effect which is contrary to its epistemological meaning.

Much of the Marxist literature which uses the realist method has also failed to stipulate the dialectical relationship between the abstract and the concrete. Keat & Urry (1975), for example, refer only to the dialectical nature of necessary relationships (Sayer 1982a) in the sense that the two parties are in opposition. For example capital and labour are not only mutually necessary entities, but they also oppose each other by definition. However, the concrete is also contradictory because events have the effect of changing the conditions in which mechanisms operate, and thus by their very existence give rise to new and different events. This is vital to an understanding of history because it emphasizes the need to consider both abstract, yet real, causal mechanisms and the evaluation of concrete conditions in explaining the actions of people which constitute events.

We have tried to illustrate the use of these epistemological principles in the execution of empirical work in two sections about the crisis in North American manufacturing industry. The first of these sections employs abstract theoretical arguments to explain broad movements in the Canadian manufacturing sector in the postwar period. Such analysis was shown to provide a useful indication of the efficacy of certain theoretical claims, but it is not particularly helpful in understanding the pressures facing individual industries, firms or people.

We would not go so far as to say that the fourth section completes the explanation. It merely shows the direction in which analysis must be extended in order to improve the explanation. It is certainly limited in that it provides no explanation at all of the crisis in the US economy as a whole. In this sense its specificity is too great to provide an answer to more than limited questions, such as those about steel industry development. But its strength lies in the combination of abstract causes, of crisis theory, and detail of specific events such as the actions of individual firms and political groups. Furthermore the historical content allows the identification of a number of changes in specific conditions surrounding accumulation in the steel industry and the difference that this makes to the influence of causal mechanisms on the events we see.

What this section cannot do is provide us with a test of Marxist theory in the conventional sense. This is because the level of detail is too great for us to specify an expected relationship between abstract cause and the form of outcome; this system is open. Class struggle and inter-capitalist competition cause such a variety of outcomes at different times and in different places, that no relationship can be meaningfully tested by conventional empirical methods. But we can attempt to provide explanations which include both determinacy and specificity (Jessop 1982) and which must then stand both on the success of the political agencies suggested by them, and on their ability to reveal the weaknesses in other attempts such as their usefulness in academic discourse. This chapter is designed as a contribution to the academic discourse about the scientific nature of Marxist geography.

7

Structure and agency in economic geography and theories of economic value

TREVOR J. BARNES

INTRODUCTION

It is unclear where economic geography fits into the recent discussions about agency and structure in human geography. Certainly there is no shortage of clues, hints and possibilities (Gregory 1981, 1982). But when it comes down to it, there are few actual proposals. One exception is the recent work of Sayer (1982a). In a brief discussion, Sayer (1982a, pp.80–82) recognizes that although economic geography has a theory of agency and structure, it has nothing in between; nothing to connect agency and structure. Thus economic geography has a voluntarist account of human behaviour (human agency) represented by the geographically free-wheeling entrepreneur who can locate anywhere, and a structural account where actors become the dupes of the geographical structure of the economy. But as Sayer recognizes, neither the voluntarist nor the structural account is satisfactory on its own. What is required is somehow to bridge the gap between agency and structure. For Sayer, at least, that bridge is realist philosophy.

My chapter has two purposes. The first is to add substance to Sayer's claims. Specifically, I will argue that the agency/structure dualism in economic geography results from the adoption of one of two theories of value found in economics. The first of these is the labour theory of value. Here goods exchange roughly in proportion to the labour time contained within them. Such a view is associated with the classical economists, although perhaps its best-known advocate is Marx. The second theory of value is based on utility. In this scheme goods exchange in proportion to their usefulness. And their usefulness is determined by both the relative scarcity of goods and their worth according to the subjective preferences of consumers. Although the use of this hedonistic calculus can be traced back at least to the time of Bentham, it was not until the 1870s with the marriage of differential calculus with utility (and their consequent child, marginal utility) that the neoclassical school, as it became known, fully came into its own. I will argue in the first part of this chapter that the use of the labour theory of value implies a structural account of economic behaviour, whereas utility theory leads to a voluntarist account.

The second purpose of the chapter is to argue that no kind of philosophical 'ism', including Sayer's own 'realism', can permanently resolve the dualism in economic geography between agency and structure. The argument here is that most of the kinds of philosophical 'isms' employed by geographers lead geographers away from actually doing economic geography. Rather than approaching the geographical world on its own terms, economic geography is reduced to whatever 'ism' is employed. Thus even if Sayer's 'realism' can bridge the gap between agency and structure, which is itself unclear, all we have done is to replace one kind of 'ism' with a different kind. Rather than simply playing off one philosophy against another, it is argued that there is another way. Specifically, it is suggested that because the problem of agency and structure in economic geography is a result of the adoption of particular theories of economic value, then what is required as a solution is a non-theory of value. This non-theory of value is discussed in relation to the work of Piero Sraffa.

MARX'S LABOUR THEORY OF VALUE

Marx's labour theory of value was contentious from the outset (Dobb 1973, pp.141–2). Rather than examining its problems, however, I intend to examine why users of the labour theory of value are drawn to a structuralist account of economic behaviour. This does not imply that all Marxists are structuralists. As has often been noted, there is clear tension within Marx's own account between structural and other types of explanation (Gouldner 1980, Thompson 1978). None the less, among Marxist economists, and more pertinently among Marxist economic geographers, structural accounts dominate (for documentation see Duncan & Ley 1982).

As the economic anthropologist Stephen Gudeman (1982) suggests, the labour theory of value is a core code for Marx. By that Gudeman means that labour values are the linchpin for Marx's whole model of economy and society. It is not that labour values are simply a useful way to examine economy and society, but rather that they define what economy and society are.

The significance of labour values for Marxist economic geographers is that they are rooted in an exogenous constant – human physical labour. This exogenous quality is crucial because whatever is used to explain the economic system must lie outside it. If it did not, the entity doing the explaining would be changed by the thing that is explained. Thus, as Dobb argues, the central requirement of any theory of value (including the labour theory of value) is:

> . . .that it is some quantity which in any particular case can be known *independently* of any of the other variables in the system. It must be something which can be postulated independently of the rest. It is some quantity brought in, as it were, from outside the system of events. . . and in an important sense it is on this outside factor that the total situation is made to depend (Dobb 1940, p.6; original emphasis).

Labour values fit the requirement of being exogenous because '(as) used by. . .Marx the concept of labour was an objective one; labour being conceived as the expenditure of a given quantum of human energy' (1940, p.13).

Given that there are other measures that fit the general requirement of a theory of value, the obvious question is why labour time is chosen by Marx and not some other variable? There have been a number of reasons put forward to explain this choice (Wright 1979), but perhaps the one that is most popular and enduring has to do with the relationship between humans and nature (Burgess 1976, Dobb 1973, pp.143–6, Lippi 1979). The argument goes that the human lot in life is to struggle with, and wrest a living from, nature. The only way in which nature is ultimately subordinated, and transformed to conform to the purposes of humans, is through the act of labour. It is for this reason then that labour time is the basis on which a good's value is calculated.

Given the centrality of labour values, it follows that the nature of economy and society within the Marxist scheme comes to be shaped by this one external constant. This will be shown by examining some key features of the Marxist conception of the economy, and of its vision of society.

What distinguishes the Marxist view of the economy from other accounts is first, the emphasis on production, and in particular the costs of production, and second, its concern with distribution. That production costs are critical, rather than, for example, demand considerations, stems directly from the broader philosophical justification of labour values alluded to above. For the central fact about commodities is that they have a direct human physical cost. Without the sweat of the brow of the labourer there would be no commodities at all. Certainly, workers may use tools, such as lathes and ploughs, to assist in production. But tools are themselves nothing but embodiments of 'dead labour', and so their labour cost too must be included when calculating the value of any good. In short, production and production costs are central to the Marxist scheme because goods are valuable only in so far as they are the *product* and *cost* of labour.

Turning now to the importance of distribution, we find that labour values have two roles. First, they provide an unambiguous measure of the distributional shares accruing to each social class. Although the issue is a complex one (see Dobb 1973), labour values enable easy reckoning of the national income irrespective of how distributional shares (measured in labour values) change. Second, using labour values implies fixed limits to the size of shares making up the national product. The important consequence is that whatever one class gains in terms of its share of national product, another class loses. This antagonistic relationship in capitalism is measured by the rate of exploitation. It is the difference between the number of hours a labourer works, and the wage s/he receives, also measured in labour values.

Having examined the economy, let us now look at Marx's vision of society. Society for Marx is composed of a number of classes. Although a controversial topic, the definition of, and relationship among, social classes depends on economic considerations. As Keat & Urry argue: 'classes are themselves functions of the process of production. Class relationships are for Marx generated within and by the fundamental structure of the capitalist mode' (Keat & Urry 1982, p.107).

What defines a social class is its relationship to production. In particular, as Cutler *et al.* argue, 'economic class-relationships involve definite forms of possession of and separation from the means of production on the part of economic agents' (Cutler *et al.* 1977, p.239). Workers are workers because all they have to sell is their own labour; capitalists are capitalists because they own and supply the basic means of production, capital; and landlords are landlords because they own a necessary non-produced input to sell to capitalists. Second, the relationships among social classes are defined in the Marxist scheme by the sphere of distribution. As suggested, what emerges from distribution is antagonism among classes, because whatever one group gains another group loses. Specifically, within the context of capitalism, because of the unequal ownership of the means of production, one class, capitalists, always ends up exploiting another class, workers. Although this is the principal antagonism, there are also other conflicts such as those between workers and landlords and landlords and capitalists.

This is obviously a very brief sketch and cannot do justice to all the complexities and subtleties of Marx's work. None the less, it is sufficient to show why, by beginning with labour values, one is led to a structural account of society. Labour values provide a charter for making production and distribution the central components in the economy. The economy in turn shapes the nature of society. Social classes are defined by their access to the means of production, and the antagonistic relationship among social classes is given by the nature of the sphere of distribution. What we are left with is a clear social structure; a structure that has the four features that Keat & Urry (1982, pp.119–23) recognize as being common to all structuralist approaches. First, there is an emphasis on the *relationships among elements*, rather than focusing on the elements themselves. In the case of capitalism, what is most important is the antagonistic relationship between workers and capitalists. And more generally, it is the class conflict endemic to all modes of production barring the communist one. Second, the structure is regular, orderly, systematic and enduring. As we have seen, the principal antagonism between social classes holds across different societies and different epochs. Master/slave, lord/serf and capitalist/worker are all variants on the same enduring structure. Third, the structure is abstract and can only be grasped when the code in which it is expressed is deciphered. Thus only when Marx realized the importance of labour values, which lie beneath the surface of prices, was he able to get behind the veil of appearances to 'really' see what was going on. Finally, the structure as a whole determines the roles people play within the system. The result is that individuals need not be referred to. Individuals simply become bearers of the social roles implied by the economy, and more fundamentally by labour values themselves. This is most clearly seen in Althusser's work:

The structure of the relations of production determine the *places* and *functions* occupied and adopted by the agents of production, who are never anything more than occupants of these places, insofar as they are 'supports' (Träger) of these functions. The true subjects. . .are not. . .'concrete individuals' – 'real men' – but the definition and distribution of these places and functions (Althusser & Balibar 1970, p.180, original emphasis).

NEOCLASSICAL ECONOMICS

Around the 1870s the key code was changed in economics. Labour values were out and utility was in. The metaphor was changed from the body, where aching muscles determined worth, to the mind, where mental desires and wants determined value. In the shuffle what was lost was a concern for social structure, and what took its place was an emphasis on the autonomous individual.

As with the case of labour values, utility is the starting point for the neoclassical vision of economy and society. The fact, as Joan Robinson (1962, p.48) notes, that there are severe problems in defining the term misses the point about utility's role. Like labour values, the usefulness of utility is not in describing the world but in saying how it ought to be.

To be recognized as a legitimate theory of value, utility needs to be rooted in a stable exogenous constant in just the same way that labour values are. The anchor that secures utility is the human mind. As Dobb writes:

To be sufficient anchorage for a determinate theory of value, even formally viewed, it was necessary that utility should be conceived as an expression of some fairly permanent and consistent aspect of human psychology. . .In so far as utility could be hedonistically treated as a fundamental 'satisfaction', then. . .it could reasonably be held to fulfil this condition (Dobb 1940, pp.156–7).

With utility tied to a firm bedrock, utility then becomes the key to explaining the economic system, and for that matter most other areas of social interchange as well. Specifically, because utility is thought to be a 'good thing', and in fact the only thing worth striving for, all human action is dependent on, and thus explainable in terms of, the adding and subtracting of utility. Furthermore, utility provides a purpose to people's lives. The ultimate goal in life is to obtain as much utility as possible because it is only utility that makes humans tick. Thus, like the teleological element in Marx's conception of society – the ultimate goal is a communist society, in which labour values are finally realized as being the true measure of worth (Dumont 1977) – there is also a teleological strain to neoclassical economics' conception of the individual; the purpose of an individual's life is to maximize utility (see Mirowski 1984).

The consequence of making utility central is to redraw dramatically the boundaries of the economic problem, and society's relationship to it, as it was conceived by Marx. To examine this issue further I will first discuss the nature of the neoclassical economy, and then look at the type of society associated with it.

If for Marx production and distribution are the two main elements of the economy, then for neoclassical economics it is clearly consumption and exchange that are critical. As already suggested, actors in the neoclassical world are impelled to maximize utility. The origin of utility is, for the most part, the consumption of goods and services. Whereas labouring is bad – a source of disutility – consumption is good; it provides happiness and satisfaction to all who engage in it. Although it is usual to think of consumption in narrow terms – goods and services that are bought commercially – one can also view consumption in a broader way. Any act that involves choices can be thought of as an act of consumption. From getting married to deciding when to wake up in the morning, humans are continually making consumption choices. That utility theory can deal with such a wide variety of situations is a result of its being rooted in an eternal constant, the human psyche. The second component in the neoclassical scheme, which is also informed by utility, is the sphere of exchange, or price determination. Because utility derives from consumption goods bought and sold in the market, it is possible to specify a relationship between prices and quantities of goods and consumers' utilities taken at the margin. Specifically, utility is maximized when the ratio of marginal utilities to prices is equal. So the neoclassical view of exchange is characterized by a one-way avenue leading from individual consumer utility functions to the prices of finished goods. It should also be noted that the way in which exchange is cast in the neoclassical model rules out exploitation. This is so because value is not determined by an outside measure of productiveness: if a good or factor service acquires a price at the market place, it by definition has its correct value. Value is determined not from where the good or service originates in production but where it eventuates in exchange.

If we now turn to the neoclassical conception of society we find, as in the Marxist scheme, that economic considerations are crucial in defining both the nature of the said actors, and the relationship between them. Specifically, the neoclassical focus on consumption leads to the individual being the basic unit of analysis, and the sphere of exchange determines the relationship among those individuals.

There are three reasons why an emphasis on consumption leads neoclassical economics to an individualistic conception of the economy. First, conceiving of consumption as a source of mental satisfaction implies an emphasis on the individual, because only individuals have minds. Social aggregates such as institutions, classes and societies do not. Second, the satisfaction originating from consumption is not inter-subjective (Robbins 1935). Although I know how much I like a good, no one else can know. As a result there can never be a 'culture of consumption' because there are never any shared meanings among members of society. Finally, to ensure that utility is maximized, social influences must be assumed away to allow the consumer to be free from outside influences that may corrupt his/her consumption choices. What neoclassical economics proposes then is a version of methodological individualism, a version that Boland (1982, Ch. 2) calls reductive psychologistic individualism. It is a view in which all economic acts, whether individual, institutional, or social are ultimately reduced to the psychological states of individuals. This position is inevitable given that for neoclassical economics, human consciousness is the key exogenous constant.

The relationship among individuals in the neoclassical scheme is given by the sphere of exchange. The problem with the reductive psychologistic individualism that neoclassical economics adopts is that there is nothing to connect individuals. Society could easily be a series of Robinson Crusoes. It is here that the market place, and thereby exchange, is critical. In their desire to obtain utility, individuals are forced to interact with one another. Through selling factor services or buying consumption goods, autonomous individuals are impelled to come together at the market place. The exchange process, in one sense, is the glue that makes neoclassical society stick together. The precise nature of the relationship among individuals at the market is one of competition, to get the most for the least. This is not the same as the antagonistic relationship in the Marxist scheme. As has been suggested, with value subjectively defined and locked away in the consumer's head, there is no external measure to show whether someone is swindled in the exchange process.

Again this account cannot do justice to all the nuances of the neoclassical scheme. But it is sufficient to suggest why, by beginning with utility, neoclassical economics is led to a voluntarist account of human action. As Giddens notes, the notion of voluntarism originates in Talcott Parsons' work where it 'refers primarily to the purposive character of human conduct, and to the capability of actors to choose between different goals or projects' (Giddens 1979, pp.253–4). Parsons, in turn, synthesized this notion from an examination of four 19th-century social theorists, two of whom were Alfred Marshall and Vilfredo Pareto (Parsons 1949, Chs 4–7). As Marshall and Pareto were both extremely influential in establishing the neoclassical research agenda, it is not surprising that voluntarism describes the neoclassical approach to economy and society. With utility defining the individual as the basic unit of analysis, society becomes simply the sum of the individuals who comprise it. Each of these individuals in turn has one purpose in life: to maximize utility through consumption. This is achieved only by making the correct consumption choices in the sphere of exchange. Thus the hallmarks of voluntarism – purposeful behaviour and constant decision making among alternative choices – are precisely those of neoclassical economics. So in contrast to the Marxian analysis that has a theory of social structure but not of individual agency, neoclassical economics presents a theory of individual agency but says very little about social structure.

ARCHITECTONIC AND CONTEXTUAL APPROACHES

The argument so far has been that the split between a voluntarist and a structural account in economic geography is a result of making utility and labour values the core code for neoclassical and Marxian economics respectively. In general terms, the use of labour values and utility represents a case of what Curry (1980) terms an architectonic approach. That is, by beginning with a set of core concepts – labour values and utility in this case – a complete system is constructed. It should be noted that this architectonic approach is not confined to economic geography but is found throughout human geography. Thus most of the kinds of philosophical 'isms' put forward by human geographers are attempts to lay a foundation for a 'complete geography'; that is, to compile a series of definitive definitions of what geography is and should be.

Now, as already intimated, Sayer's solution to the agency/structure dualism in economic geography is realist philosophy. But as will be argued, it is not clear in what sense Sayer's solution is a real solution. For in advocating 'realism' all that Sayer does is to swap one core code for another, the core code underlying voluntarism or structuralism for the core code of realism. To put this another way, Sayer's argument leads to batting around the question of which architectonic approach is best. But this misses the point, a point that Sayer might have realized if he had enquired how the agency/structure dualism in economic geography arose in the first place. For the problem is not which architectonic approach is best, but rather with the whole architectonic approach itself. In particular, there are three problems with any architectonic approach. First, in building 'complete systems' geographers are often trying to solve problems that confound philosophers. Although it is possible that geographers will solve hitherto unsolved philosophical quandaries, the more likely result is confusion. Second, by adhering to one of these prior philosophies, geographers end up not doing geography, but reducing the geographical world to the philosophical 'ism' employed. In effect, the geographical world is philosophized away. And finally, there are no logical criteria by which to decide which philosophical 'ism' is best, because it is precisely by means of that 'ism' that what count as logical criteria are decided.

Any proposed solution to the agency and structure dualism in economic geography that argues on the same terms as that dualism, that is, by comparing one architectonic approach to another, ultimately fails. For such a solution is pitched at the wrong level. It supposes, as Sayer did, that the problem is to select among sets of core concepts, whereas the real problem is to avoid using core concepts to begin with.

What then is the alternative? I will argue it is one based on context. Unlike the architectonic approach that supposes that there are only certain ways of knowing and only certain independent objects to which knowing applies, a contextual approach suggests that there is no 'real' knowledge and no single way that knowledge can be acquired. We simply cannot specify a priori any absolutes about knowing. Rather, in order to know what knowledge, understanding and objectivity are, we must look at how they are used, acquired and verified in the context at hand. Whereas the architectonic approach squeezes the world into pre-formed philosophical categories, a contextual approach begins with the world intact. It celebrates the richness and diversity of the world, and does not reduce it to something it is not (Barnes & Curry 1983).

As an example of a contextual approach, one that economic geographers may draw upon, the work of the economist Piero Sraffa (1960) will be presented. In contrast to the theories of value already discussed that are rooted in the supposed universals of labour value and utility, Sraffa provides a non-value theory; a value theory not rooted in anything, and thus able to be embedded in any context.

SRAFFA'S NON-VALUE THEORY

Sraffa's publications give no explicit indication of his philosophical position. In working out that position, one can examine the context in which Sraffa wrote[1] or one can examine the written texts themselves. Of course, best of all would be to do both (see LaCapra 1983, on this point), but for brevity's sake I will concentrate on the second alternative and examine Sraffa's only book *Production of commodities by means of commodities* (1960).

Sraffa's book is unlike any other in economics. There is no introduction, nor conclusion; there is no mention of individual economic actors nor social institutions. All we are given are the equations of production. Although some have interpreted this terseness as simply a stylistic quirk (Harcourt 1982, pp.271–2), another interpretation is that what Sraffa leaves out is more important than what he includes. This is true in two ways. The first is evident when we recall the book's sub-title: *Prelude to a critique of economic theory*. Sraffa is not offering the critique itself, but rather offering the wherewithal for others to carry out the critique. Specifically, *Production of commodities by means of commodities* implicitly contains a logical critique of both neoclassical value theory and Marx's labour theory of value.[2]

The second issue is that Sraffa's reticence suggests that he realizes that there are definite limits to economic discourse. These limits are not stated explicitly; rather, they have to be surmised from what Sraffa leaves out of his discussion. To understand these limits, let us turn to the substance of his work.

Sraffa portrays an economy where inputs, that are themselves commodities, produce other commodities. The process is a circular one where the output in one production period is used as an input for the next production period. Sraffa begins his discussion with the case of a subsistence economy. What is produced at the end of the production period is just sufficient to replace the inputs used up. In most economies, however, there is a surplus left over even when all inputs are replaced. Thus we might imagine that in a feudal economy most of the surplus goes to the lord of the manor, whereas in modern capitalism it is divided between workers and capitalists. What Sraffa then demonstrates is that regardless of whether we have simple reproduction, or reproduction with a surplus, it is always possible to derive a unique system of exchange ratios that will keep the system going.

An immediate thing to note about Sraffa's model is that there are no absolute sources of value that are determining prices. Prices of commodities are determined by the means of production, and the means of production are themselves composed of commodities that are produced by a previous set of commodities. At every stage then commodities are produced by other commodities, and as a result there is never an ultimate source of value.

Sraffa has been discussed by both neoclassical and Marxian economists. From the neoclassical camp Sraffa's work on production is seen as nothing more than a special case of the neoclassical model (Hahn 1983), whereas some Marxists seem to think that all Sraffa provides is a mathematically correct solution to the problem of transforming labour values into prices (Meek 1976). But both views miss the point. Whereas neoclassical economics is about the production of goods defined in terms of utility and Marxian economics is about the production of goods defined in terms of labour values, Sraffa is only concerned with the production of 'things'. The word 'thing' is used deliberately because it implies no valuation; 'things' only take on meaning when they are embedded in a particular context, and that is Sraffa's whole point. Napoleoni also recognizes Sraffa's concern with 'things' when he writes: 'A re-reading of *Production of commodities by means of commodities* has confirmed my initial impression that it is things rather than subjects which move around in Sraffa's construction' (Napoleoni 1978, p.75). This leads Napoleoni to suggest that what Sraffa leaves us with is a *tabula rasa*. It is in this sense that Sraffa presents a non-value theory. We must wait to learn how things are valued in different places and at different times; we must look at the context.

As there are no ultimate values in Sraffa's system, no form of behaviour is prescribed. The production of commodities by means of commodities must of course be sufficient to reproduce the system, but once this viability constraint is satisfied, Sraffa does not say any more. Sraffa does not lay out a theory of value, nor does he lay out a theory of distribution once surplus is obtained. The values of goods and the way in which they are distributed depend on the particular context in which the

production system is embedded. Because Sraffa leaves us with a
tabula rasa some economists are frustrated. They want a cause for
value and a mechanism to explain distribution. But again this
misses the point of Sraffa's work. As Napoleoni argues: 'Sraffa
forces us to take everything back to the beginning. True, he gives
no suggestion as to how we should begin again: he would not, I
think, wish to give any; and I am not sure he would be able to'
(Napoleoni 1978, p.77).

We are now in a better position to understand what Sraffa
thinks we can say about an economy and what we must be silent
about. At the most general level, Sraffa argues that for there to be
continued economic activity, the output of commodities must
exceed or at least be equal to inputs of commodities. About the
meaning of these commodities, the manner in which income is
distributed, the level of consumption, the social institutions
present, and the nature of the economic actors, Sraffa says nothing
because nothing can be said a priori.

It is true that when Sraffa discusses an economy with surplus
product, he is clearly discussing a capitalist economy – he uses
terms such as wages and rents, and assumes profit equalization.
However, to conclude from this that Sraffa is now offering a
closed rather than an open system misses the distinction between
the specific and the general. Although wages, rents and profits are
the distributional variables Sraffa discusses, they could in fact be
anything from tribute paid to the lord of the manor to potlatch
paid to the Big Men (see C. Gregory 1982; Gudeman 1978).
Wages, rents and profits are simply terms used to designate
portions of the surplus product. As Sraffa does not provide a
theory of profit and wage determination, nor discuss who receives
these shares of the surplus product, they might equally well take
the form of some other distributional variables. What form they
take will depend on the context in which the economy
investigated is embedded.[3]

Having discussed Sraffa's work in general terms, let us now
turn to its relationship to the agency/structure dualism. In one
sense, Sraffa's ideas can be seen as undermining the usefulness of
this distinction altogether. Because Sraffa does not have a theory
of value that prescribes a priori a particular method of treating
individuals and society, the question of whether Sraffa is
proposing a structuralist or voluntarist account is irrelevant. It
may well be that in one context it makes sense to employ a
structuralist approach, while in another context it is best to
emphasize agency. Or then again, on another occasion it may not
be useful to refer to structure and agency at all, while on yet a

different occasion one may want to refer to both. The point is that Sraffa's work is compatible with all of these approaches because the Sraffan account is not a 'complete system' that has answers to every question. Only by examining the context at hand can we know how structure and agency fit in, if at all.

One final point to note about the contextual position implied in Sraffa's work is that it insists that in order either to undertake research or to understand someone else's research we must see such work as arising within a particular intellectual context. A contextual position not only makes us sensitive to the particular context studied, but it also makes us sensitive to the context in which the person undertaking the study is part. For a researcher always has certain aims, motivations, audiences and methods (Curry 1985). These form part of the intellectual context in which the study is carried out.

The consequence is that studies often reveal as much about ourselves as about the thing we are studying. For example the interest in structuralism in urban geography in the late 1970s tells us almost as much about the disillusioned positivists undertaking these studies, the frenzy of experimentation in human geography at the time and the re-emergence of Marxism in social science in general, as it does about the urban phenomena studied. Thus a contextual approach not only challenges us to find an appropriate way of dealing with the specific context, but it also challenges us to think about the way we think about that context.

CONCLUSION

In conclusion, this chapter has argued that the dualism in economic geography between agency and structure is a result of the adoption of two particular theories of value, namely, the labour theory of value and utility theory. More generally, this agency/structure dualism is a result of the acceptance of the architectonic approach, an approach concerned with building a 'complete system' from a limited set of core concepts. There are, however, a number of problems with the architectonic approach. This means that any proposed solution to the agency/structure dualism, such as Sayer's, that still relies on an architectonic approach ultimately fails. An alternative to the architectonic approach, one that can truly deal with the agency/structure dualism, is one based on context. The work of the economist Piero Sraffa was presented as an example of such a contextual approach.

One final caveat should be made. Advocating a contextual approach does not mean that substantive conclusions of structuralism, voluntarism and realism do not have a place in human geography. But their place must be discovered within particular contexts and not assumed a priori.

NOTES

1 There is evidence that links Sraffa with Wittgenstein; see Fann 1971, McGuiness 1982, Roncaglia 1978.
2 The former is elaborated upon in the 'capital controversy' (Harcourt 1972); the latter is discussed in the 'value controversy' (Steedman & Sweezy 1981).
3 The Panamanian and Papua New Guinean tribal economies have recently been analysed from a Sraffan perspective; see respectively Gudeman 1978 and C. Gregory 1982.

8

Responsive methods, geographical imagination and the study of landscapes

EDWARD RELPH

INTRODUCTION

There are methods for the study of landscapes and places that are both systematic and responsive. Unfortunately the methods that are systematic are insensitive, and the methods that are responsive cannot be reduced to simple handbook procedures.

The problem is this. Landscapes and places are the contexts of daily life, they are with us as we walk down the street or look through the windshield; we manipulate them in gardens, add to them with our purchases, use them as residents and tourists, and give meaning to them in our imaginations and memories. They are, in short, vital and subtle things, filled with specific significance and incapable of exact delimitation or definition. Scientific and systematic methods, on the other hand, are standardized, objective, capable of being used by others and producing results that can be checked. Their essential merit is that they make a complex situation intelligible by imposing an abstract framework on it. Trying to investigate places and landscapes by imposing standardized methods is like studying ballet by putting the dancers in straitjackets, or judging wines by measuring their alcohol content – the information obtained may be accurate but it seriously misrepresents the subject matter.

This comparison is not exaggerated. There are people who measure what they call landscape quality and attachment to place by means of a complex apparatus of photographs divided into grids, psychological tests, questionnaires and statistical analyses. The results of all this have the same mathematical form as a measurement of alcohol content and are therefore enthusiastically received by bureaucrats and others who prefer a whole number to a clear thought, but as the entire procedure is based on arbitrary definitions of the landscapes and places being studied (just where do you draw a boundary around landscape?) the results can in fact mean anything one chooses. This must be the sort of approach the philosopher Ludwig Wittgenstein was referring to when he wrote of psychology that it has 'experimental methods and conceptual confusion' (Wittgenstein 1958, p.xiv).

Systematic and scientific methods are imposed on a subject matter, wringing information from it. Responsive methods – they might also appropriately be called phenomenological or artistic – are adapted to the form and character of the subject. Actually, when put this bluntly the distinction between the two is overdrawn, and elsewhere in this chapter I equivocate about it, but in general I have no doubts that it is the responsive methods which are suitable for investigating and teaching about landscapes, human geography, or indeed any phenomena in which particularity and meaning are paramount. Such phenomena have to be approached with the sensitivity and imagination that are commonly found in those methods of art that reflect uniqueness and cannot be reduced to universal sets of procedures. This might seem to imply that there is nothing in these methods that can be clarified or practised, that they are egoistic and their results are tritely subjective. I accept none of these charges. Responsive methods can be clarified, though not reduced to handbook knowledge; they can be rigorously practised; and I consider them to be an alternative and an antidote to the programmatic and often destructive methods which are widely used for the analysis and manipulation of human environments.

In this chapter I attempt to clarify and describe the use of these responsive, phenomenological or artistic methods as I understand them and use them both in my research and my teaching (I do not draw a neat line between these two activities). In the interests of brevity and clarity I will direct my remarks to the study of landscapes; places are, I think, rather more complex phenomena but do not require significantly different approaches. I begin at a philosophical level; this reflects my conviction that method in the absence of philosophy opens the door for confusion and even violence because it is detached from its logical and ethical contexts. I then consider the practicable aspects or responsive methods, first by reference to art, and second by a discussion of the importance and use of the geographical imagination as a way of keeping topics in perspective.

A PRELIMINARY CAUTION – A SOCIETY OF THE LANDSCAPE-BLIND

One of my worst imaginings is of a society in which most people drift around in a detached haze of abstractions and borrowed opinions, noticing nothing of their surroundings for themselves. They know of few places, or ways to get to them, on the basis of their own experience, but rely on street guides, the suggestions of others and directional signs. The only words they have to describe the world around them are vague, abstract ones like housing, dwelling unit, central business district, retail facility, or recreational resource; their only aesthetic responses are to cliché picturesque scenes designated for them by markers or historical plaques; they mistrust the evidence of their own eyes and thoughts but regard the opinions of remote experts as sacrosanct.

In all previous cultures everyone needed some direct knowledge of their environments in order to survive. The fact now is that many people, at least among those who live in cities, can exist with a minimal direct experience of their surroundings, relying on packaged information or second-hand advice. This sort of detachment from the world one lives in should not be encouraged. First of all, it opens the way for the political and commercial manipulation of attitudes. Second, it is my conviction that landscapes and places matter. They are not just incidental visual backgrounds to other social concerns but are part of our being that enter directly into the quality of our lives by providing countless small pleasures, by stimulating our thinking, and even by providing the context for moments of profound revelation. If the abilities to attend for oneself to the qualities of landscapes and environments are missing, then life is greatly impoverished. Accordingly, much of what I do in teaching and writing is generally directed toward the humanist aim of achieving a reasonable – I use the word in its exact meaning – independence in observation, thought and judgement, both for myself and for my students, and is more specifically directed toward an appreciation of landscapes and places. For me these are the essential purposes of human geography.

SEEING, THINKING, DESCRIBING: RUSKIN, HEIDEGGER, WITTGENSTEIN

Investigating actual landscapes requires three closely related activities: observing or seeing, thinking about what has been seen and description. I look carefully at the landscapes I encounter, I reflect on why they are as they are and what they disclose to me and I then attempt to give a straightforward account of what I have seen and thought (Relph 1984). These activities are rather more difficult in practice than this simple summary suggests, and for advice and direction I find myself turning repeatedly to the writings of John Ruskin, Martin Heidegger and Ludwig Wittgenstein.

Ruskin was a 19th-century art critic with enthusiasms both for landscapes and for seeing clearly. He cultivated his ability to observe carefully and precisely by making studies of clouds, mountains, geological strata, trees and buildings. He then wrote eloquently about the necessity for and practice of clear seeing. He chastised artists who saw as if in a state of visual confusion, painting either minute details far beyond the range of human vision, or generalized trees and animals. 'It is just as impossible to generalise granite and slate,' he wrote 'as it is to generalise a man and a cow. An animal must be either one animal or another animal: it cannot be a general animal or it is no animal' (Ruskin 1856, s.28).

Heidegger's philosophical writings are a phenomenological exposition of Being, of the experience of existence. His work is a demonstration, and sometimes an explicit discussion, of how to think about subtle questions of meaning and responsibility. From Heidegger I learn and relearn the possibilities offered by phenomenological ways of thinking which do not impose methodological systems but instead respond to their subject matters and to experiences of these.

My understanding of description derives in part from Wittgenstein's philosophical accounts of language. He concluded that the most important aspects of life cannot be explained scientifically; all one can do is describe them in their context and that might clarify their meanings. This suggestion is an important one in the context of this chapter, and warrants elucidation. The word 'description' can mean many things, including what is usually understood to be interpretation, and indeed in so far as any description is selective it must offer a point of view. I understand description and interpretation to differ only in emphasis – description is rather more concerned with presenting observable facts, for instance of a landscape, whereas interpretation stresses some perspective or context which suggests the meaning of these facts. In both cases the chief aim is to put together pieces of information in a way that makes sense; it is of course the case that somebody else might put together those same pieces in a different way to convey a different sense.

Description and interpretation differ from explanation because explanation seeks causal mechanisms and processes, and digs beneath the surface appearance of things to try to establish what these causes are and how they work. Because the primary feature of landscapes is their surface appearance, trying to explain landscapes in this way is no better than a biologist killing an animal to find out what makes it live (see Evernden 1985).

In practice the three activities of seeing, thinking and describing rarely occur in an orderly sequence. Observations in an unfamiliar city can stimulate new ideas, or the process of writing may raise questions that stimulate thinking and the need for further looking, and so on. They are mutual and continually repeated activities which I undertake with reference to the landscapes I live in or encounter for a variety of reasons, including travel, field trips and chance. I do not self-consciously and systematically select specific landscapes for research investigations.

LANDSCAPES AND THEIR SIGNIFICANCE

Landscapes are many-sided phenomena. In their most obvious sense they are visible environments which not only have aesthetic and symbolic properties but also serve many purposes, some of which are immediately apparent, like car parks and tree-houses, while others are more subtle, such as property speculation. They can also be an entry point for addressing social and political issues, because they are as much a source of information about social conditions as documents, interviews or censuses. Considerable expenditures of time, effort and money are required to make and modify built environments so it is unlikely that their appearances are mere accidents. Even mess and dereliction inform us about values and priorities; the abandoned buildings of the Bronx or the rational order of a public housing project tell us no less about the values of contemporary society than the skyscrapers of Lower Manhattan.

Landscapes provide me with something that I can observe directly, and to that extent my work has more in common with a physical scientist than with a social scientist grappling with the chimeras of markets and class systems. On the other hand there is no scope for experiments, and I do not try to explain landscapes in terms of some theory or model, or to give measures to their qualities. Like a juror weighing evidence, I attempt only to give an accurate interpretation of what I have seen, read and thought about, and to allow the form of this account to reflect the character of the landscapes with which I am concerned. In much of my writing I have not been concerned with particular landscapes only for themselves, but have attempted to look through them to identify underlying processes and patterns of thought, such as the placelessness and rationalism which infuse most modern environments.

What is the practical value of such landscape studies? Note first of all that little of this fits with the currently prevailing paradigm of how the world works, which is as a great machine made up of lesser intermeshing machines usually called systems – ecosystems, social systems, economic systems, urban systems and so on. In the context of this paradigm practical application means developing techniques to improve the efficiency of systems of operation. Responsive methods have little or nothing to contribute to such applications; they have no panaceas to offer for either small or large problems. What they provide, to borrow an apposite expression from the phenomenological psychologist Erwin Straus, is not knowledge 'meant for mastering the world, but rather for unlocking it and making a world that is mute speak to us in a thousand places' (Straus 1963, p.395). This is not unlike the specific claim of J.B. Jackson that landscape study makes for 'alert and enthusiastic tourists', though he appends to this the Socratic comment that by knowing more about the world we come to know more about ourselves (Jackson 1980, p.3).

Joe Powell, a geographer concerned with the qualities of landscapes, has written that the role of geographers is to be 'interpreters of our world: to bear witness to its variety and significance, its beauty and its horrors. . .by affirmation and protest' (Powell 1979, p.333). Writing and teaching about landscape can promote independence in thinking and judgement, can help to develop an understanding of the complex and open-ended processes that characterize all human–environment relationships and, perhaps, in the very long run and through an emphasis on responsiveness and responsibility, they might contribute to the discovery and acceptance of ways of living on the earth which are less technically violent and more sustainable than the ones which now prevail.

ARTISTIC METHODS IN THE GEOGRAPHICAL STUDY OF LANDSCAPE

Regardless of its merits it may still appear that landscape study has no methodology other than, in C.W. Mills' (1971) phrase, some type of obscure 'intellectual craftsmanship'. Appearances can be deceiving. While it is not possible to elaborate a complete methodological system, there are explicit techniques that provide a foundation for the study of landscape within geography.

Donald Meinig, writing with reference to geography, has suggested that there is a significant difference between humanism and the humanities, and that in rejecting humanism as a confused body of sentiments it is not necessary also to reject the various insights and methods that are part of the humanities. Accordingly he argues that geography can be approached as an art (Meinig, 1983). This interesting idea is worth exploring.

Is there method in art? Are we to assume that painting and writing are mysterious, magical processes involving no techniques which can be taught and learnt? Those with narrow scientific minds might be inclined to think so. For them any procedure not capable of being set out as a sequence of rules which can be applied uniformly and guaranteed to produce standardized results is both mysterious and of doubtful merit. In art there is no universal equivalent to scientific method, nevertheless artists and novelists and poets do communicate their ideas. In fact all forms of art involve techniques which can be elaborated and self-consciously practised, though these techniques often remain latent. And when they are made explicit they are usually developed in ways which, unlike the methods of science, disclose rather than deny individuality. The abstract painter Paul Klee, much of whose work consisted of repetitive experiments with line and colour in order to find the most effective ways of conveying his experiences of nature and of landscapes, was asked to give a talk about his methods. He acknowledged the force of the admonition that the painter should paint, not talk, but also found that he could indeed give words to hitherto 'nameless processes' and 'instinctive gestures' which he employed. However the purpose of these processes and gestures was, for him, not to encourage great acts of self-indulgent creativity. The artist, he wrote, 'does nothing other than gather and pass on what comes from the depths (of nature). He neither serves nor rules – he transmits. His position is humble' (Klee 1948, p.15).

Truth and accuracy are no less important for the artist than for the scientist. John Ruskin devoted much of his early adult life to demonstrating the accuracy of Turner's paintings in terms of the closeness of their observation of rocks, trees, light. These paintings were reviled by most contemporary critics who considered Turner to be wildly fanciful or to have thrown breakfast at the canvas; Ruskin demonstrated that it was the critics who were wrong, for they saw the world through clichés and contrived artistic conventions, and it was Turner who represented accurately what he saw. Nevertheless there is a difference between art and science in their concern for truth. Art usually seeks to convey the truths of the world as mediated by experience, science usually seeks the truth of the material processes of nature as though independent of human experience.[1] The paintings of Klee and Turner represent landscapes not as law-governed systems of natural processes but as expressions of continually changing experiences.

Painting and writing are not instinctive and spontaneous activities. The special talent some individuals may have for them has to be developed by training in sketching from life, in techniques of shading and brushwork, in grammar and composition, in style, etc.; this requires self-discipline, commitment and practice. The techniques of the arts and the humanities are, however, not as overt as those of science, and except for some school exercises, they are not usually treated in isolation but are learnt, practised and improved in the context of substantive work on particular topics. There is no general method for the arts and the humanities that is equivalent to scientific method, and while on occasion there have been precisely established conventions which artists have followed, for instance religious symbolism in medieval paintings or rules of perspective, these soon become sterile.

Instead of hunting for a clean distinction between art and science which can only lead to false oppositions and acrimony, I think it is more valuable to look for a distinction between good and bad forms of art and science. This too is intellectual marshland, but at least thinking about this distinction can set us on a worthwhile path toward more solid ground. Ruskin provides some hints on how to proceed. Writing about painting, he proposed that great art was characterized by its embodiment of great and noble ideas, and by its accuracy of observation and refinement of perception; bad art was poorly observed, factually incorrect, superficial and dealt with trivial subjects (Ruskin 1856, Vol. III, Ch. 2). Almost identical comments could be applied to science, with the additional qualification that bad science depends on programmatic, repetitive methods such as those used in corporate research (set 100 hypotheses, test them all and with a bit of luck one will result in a marketable product). These involve no creativity, no thoughtfulness or reflection, merely the rote following of rule-governed procedures. Such methods have economic and bureaucratic value because they are predictable. They have little intellectual merit.

All of this has several implications for understanding method in human geography and landscape studies. First of all, if one chooses to pursue geography as an art rather than as a science this still involves the use of rigorous methods, though these cannot be articulated with the same degree of formality as those of science. Second, these methods are concerned with the representation of the truths of environment, landscape and place as they are mediated by our experiences, that is with being responsive to the world around us. Third, the debate about whether geography is an art or a science has no great methodological significance and there are no firm grounds for maintaining that one is superior to the other; whether geography is approached artistically or scientifically is largely a matter of personal inclination. And finally, I believe that good geography is that which is based on direct and careful observation, which grasps relationships between personal experiences and larger patterns, which does not thought-lessly use borrowed ideas and methods, and which does not contribute to violence, exploitation or environmental degradation. There are, of course, no exact criteria for any of these, and each one of them could be argued interminably. Perhaps what is really important is the attempt to draw a distinction between poor and worthwhile geographies, to think it through as clearly as possible, and then to err in one's own work always in the direction of what is considered to be good.

THE GEOGRAPHICAL IMAGINATION

In his essay on 'Geography as an art', Donald Meinig suggests that the artist must draw upon insight and feelings and literally imagine them into forms that will make an effective connection with the lives of others (Meinig 1983). The geographical imagination is a particular means of attempting this that is especially valuable for human geography and environmental understanding; it is a way of thinking that seeks to grasp the connections between one's own experiences of particular landscapes and the larger processes of society and environment, and then seeks to interpret these in a manner that makes sense for others.

Hugh Prince once wrote that the geographical imagination is aesthetic and poetic, it gives meaning and value to the facts of place (Prince 1961, p.23). This is exactly so. It looks first at, and then through, individual experiences of places to broader environmental implications and then back to the experiences of others in order to appreciate the conditions of their lives. The geographical imagination does not stand in isolation; there are also biological, historical, sociological and other forms of imagination which seek to capture the particular concerns of these fields of knowledge. Thus C. Wright Mills noted that each of us has worries about our financial security, our family, our health and so on, and we also have various enthusiasms and convictions; these always have social context, they are framed by institutions, are part of larger social movements and contribute, albeit in a small way, to the historical processes of social change. The sociological imagination is, he suggested, the habit of mind which grasps these relationships between individual and social matters (Mills 1971). From a geographical perspective these social matters also have a landscape and an environment; our lives are necessarily framed by landscapes, and our purchases and other actions contribute to the processes of environmental exploitation and sustainability. The distinctive quality of the geographical imagination is that it aims to grasp personal, social and environmental processes in their interrelationship. For the person who has developed the geographical imagination no individual actions are without environmental and social consequences, and nowhere is remote, for the entire earth is implicated in each of its places.

The 19th-century French geographer Elisée Reclus once described humankind as 'the conscience of the earth' (Reclus 1899, p.299). This is an evocative phrase, a challenge in an age of potentially enormous environmental destruction. It implies that each one of us has a responsibility for the earth and for the environmental implications of human actions, both our own and those of others. To understand and accept this requires above all an act of geographical imagination. There can be no more important task for the art of geography than to convey this way of imagining and thinking.

WAYS OF PRACTISING RESPONSIVE METHODS

Kimon Nicolaides wrote of drawing: 'To understand theories is not enough. Much practice is necessary' (Nicolaides 1941, p.xiv).

What practices can there be for the geographical imagination and the responsive study of landscapes? All art requires perseverance and the patient acquisition of technical skills until they become almost second nature – learning scales, shading, grammar and punctuation. Similarly there are techniques that can be used to develop the skilful study of landscapes, whether in others through teaching, or in one's own practice.

1 Ways of seeing

Nicolaides claimed that 'There is only one way to draw, and that is the perfectly natural way. It has nothing to do with artifice or technique. It has nothing to do with aesthetics or conception. It has only to do with the act of correct observation, and by that I mean a physical contact with all sorts of objects through all the senses' (Nicolaides 1941, p.xiii). His book *The natural way to draw* consists of a series of exercises intended to achieve adequate drawing skills through careful observation and clear seeing through drawing.

It is not only through drawing that one's powers of observation can be developed. Raymond Williams has written that 'We learn to see a thing by learning to describe it', for description requires a kind of discovery (Williams 1965, p.39). And Samuel Johnson, in an aphorism of remarkable relevance for geography, declared that 'The use of travel is to regulate imagination by reality, and instead of thinking how things may be, to see them as they really are'.

Putting these ideas together it is apparent that describing landscapes, whether by drawing or writing about them, is a way of improving one's ability to see and to sense them clearly. Written descriptions should be brief, for brevity demands the selection of significant details, and a long account easily degenerates into an ill-assorted list of bits and pieces. They should attempt to represent the landscape accurately and honestly without resorting to a trite inventory of features; borrowed knowledge, whether from books or other sources, should be weeded out because the aim is to see for yourself and other people's ideas will get in the way of this. Judgements should only be made on the basis of careful observation. The overall aim, which sounds much easier than it is in practice, is to make a clear, concise account of somewhere so that this place would be recognizable to somebody else through your description.

The only way to write a description like this, it turns out, is to look carefully for oneself, to think about what has been seen, and then to discover an effective way of conveying these observations and thoughts. John Ruskin offered an especially useful piece of advice for doing this when he wrote that 'a curiously balanced condition of the powers of the mind is necessary to induce full admiration of any scene' (Ruskin 1857, Vol. III, s.5). Too much observation leads to a confusion of unsorted details; too much thinking imposes inappropriate ideas and interpretations.

2 *Ways of thinking*

Thinking that is responsive rather than aggressive Heidegger called 'meditative' (Heidegger 1966). Aggressive thinking (Heidegger's term for it was 'calculative') involves poor habits of thought, such as using methods developed in other contexts, or imposing arbitrary judgements. Meditative thinking offers no systems or remedies; instead 'it demands of us not to cling one-sidedly to a single idea, nor to run down a one-track course of ideas' (Heidegger 1966). Heidegger wrote at considerable length about this way of thinking, indicating beyond doubt that it cannot be summarized without serious misrepresentation. There are no technical tricks for learning meditative thinking; it is based in self-knowledge and requires a continuous effort of critical reflection that challenges any tendency to rely on second-hand techniques or habits of thought. In teaching perhaps the most that can be done is to provide circumstances conducive to such thinking. Nicolaides offers some sense of what is involved in the preface to his book. 'The job of the teacher,' he wrote, 'is to teach students, not how to draw, but how to learn to draw. They must acquire some real method of finding out facts for themselves lest they be limited for the rest of their lives to facts the instructor relates' (Nicolaides 1941, p.xiii). There is, he continues, no formula for doing this; his method is to enable students to have an experience, a deep insight, by planning for them things to do and wonder about, and then to point out what has resulted from it. After that it is up to individuals to maintain and develop this insight and their abilities to see and think for themselves.

3 *Ways of describing*

In methodological discussions, skill in writing is usually either ignored or assumed. There are, in fact, serious problems of obscure writing in all academic disciplines. Socspeak and economese are familiar forms of jargon-laden language being used to convey almost nothing verbosely. There is no equivalent term in geography, though Mark Billinge has recently identified in human geography a 'mandarin' dialect (the term is George Orwell's) which involves a flowery appeal to sentimental ideas of uncertain substance (Billinge 1983).

C.W. Mills offers simple, sensible advice: 'Present your work in as clear and simple language as your subject and your thought about it permit' (Mills 1971, p.217). He further suggests (p. 219) that to overcome academic prose you first have to overcome an academic pose by clarifying in your own mind answers to three questions. (1) How difficult and complex is the subject matter? There are no set answers to this, except that the writing should be no more obscure than its subject. (2) What status am I claiming for myself as I write? The very act of writing assumes that one has something worthwhile to say and it is therefore misleading to affect a neutral style. E.B. White suggested that 'Every writer, by the way he uses the language, reveals something of his spirit, his habits, his capacities, his bias' (Strunk & White 1979, pp.66–7). So to write prose that is deeply contorted or so precisely structured that it could have been manufactured by a machine is a serious admission about one's personality defects. (3) For whom am I trying to write? A strictly academic audience requires an attempt to clarify and inform – the aim must be to grasp and then to convey the salient features of some issue to an audience that is already committed; a certain level of specific knowledge can therefore be assumed and a relatively technical language used. For a more general audience the aim has to be to entertain as well as to inform; the interest of the readers cannot be assumed, it has to be claimed. This demands a more expressive and imaginative approach. In short, different styles have to be adopted for different audiences.

THE VALUE OF RESPONSIVE METHODS AND THE GEOGRAPHICAL IMAGINATION

An important consequence of seeing clearly and directly is that should you then try to change the world in some way, at least you will do it from knowledge which is based on your own experience. It is all too easy to follow intellectual fashions, or to work with abstract models, and then to impose these on to places and environments encountered only vicariously through a haze of words or statistics or formulae.

The uses of responsive methods as qualified by the geographical imagination are these: by teaching, to encourage and to direct others to experience landscapes and places for themselves, and so to arrive at a better self-understanding; by writing (or any other medium), to illuminate the places and environments of the world, and to interpret them, good or bad, so that others can appreciate their significance.

There are two related uses, perhaps less familiar in a field of geography widely considered impractical. First, these methods have a role to play in the disclosure of obscured truths, in undoing environmental injustice and hardship, and ultimately in finding ways to restrain those human tendencies that would destroy the earth, any of its places or any of its peoples. This implies that geographers should act as advocates for the earth and its diverse environments, possibly by adopting a role as interpreters of landscapes and places, more actively by protesting against environmentally destructive behaviour, or even by striving to set an example of environmental non-violence that others can follow. Second, the geographical imagination can contribute to the design of places and landscapes, literally by imagining alternative possibilities and arrangements for settlements and then arguing for changes according to one's convictions, and demonstrating how such changes can be effected.

Though they are enormously important there is nothing obscure about these aims. They amount simply to finding out about geographical matters through the direct study of landscapes and places, informing others of what has been discovered and then using this knowledge to argue for whatever social and environmental changes seem to be beneficial. Of course what we consider to be beneficial will depend on our personal and political convictions; I believe that these should not be argued or promoted in ways which unreasonably restrict or deny the convictions of others. And a final caution; in attempting to use the geographical imagination in these various ways it is always wise to keep in mind William Blake's forceful charge that 'He (sic) who would do good to another must do it in Minute Particulars; General Good is the plea of the scoundrel hypocrite and flatterer' (Blake n.d., *Jerusalem*, Ch. 3, Pl. 55, lines 60–63).

A CONCLUDING COMMENT

It is, no doubt, possible to say more than this about responsive methods in the art of human geography, to elaborate ways of seeing, thinking and describing and to devise techniques for enhancing these. To do this in the abstract would, however, not be very helpful; it might even be downright regressive by presenting as generally applicable a methodological framework which is inappropriate for many topics. Methods for the art of geography are better developed in the context of studies of specific issues and landscapes, and for these there can be no handbook knowledge and no unfailing guidelines. Therefore a strategy of profound good sense is to follow Joe May's advice, repeated from Voltaire's *Candide*, to bestir ourselves and 'go and work in the garden' (May 1970, p.253).

NOTES

Wayne Reeves and Richard Harris made valuable comments on a draft of this paper, and Rod Watson drew my attention to several useful articles.

1 A notable exception occurs in sub-atomic physics.

9

A critique of dialectical landscape

AUDREY KOBAYASHI

INTRODUCTION

The process of landscape is very much like that of dance. In dance, there is a long journey from the *barre* to centre stage. Along the way, the discrete *pliez* and *relevez* of the *classe* give way to undifferentiated movement. The dance cannot be reduced to its basic elements, but extends beyond itself as a totalization that includes its history and its potential. It releases time, space, form and movement, to bring dancers and audience into a common sphere of expression. And yet, we are fooled. For, the moment when the *pas de deux* reaches its exquisite climax, when the world seems to begin and end with a single subtle gesture, *that moment*, could never exist – or could never be the same – without each agonizing *pliez* that has gone before, without every dedicated encounter with the cold reality of the *barre*.

Landscape, too, is a process totalized, the extension and collection of human activity in a material setting. It is organized, choreographed, although seldom so expressively as in the dance; like the dance, it can produce profound emotional effects on the beholder whose senses are awakened to its presentation. Both landscape and the dance require a practised eye to appreciate their finer qualities. Both are irreducible to their constituent elements, both transcend their moments of expression, yet neither exceeds its physical features: buildings or props, movements, expressions in flux. Landscape too is a form of language, and it is necessary to know its history, its structures and its syntax if it is to be understood. Landscape is the assembly, the congregation, the panoply, of the dance of life.

The study of landscape provides for geographers a means of analysing and organizing the surrounding material environment, confirming the major tenet that we look to that material environment for concrete expression of the facts of human existence. Yet despite the well-established position of landscape study within the discipline there remain questions concerning how landscape is defined, and even greater questions concerning the relationship between landscapes and human beings. This issue has become more pressing in recent years in the general shift of interest from the study of landscape and culture (where culture was viewed as a phenomenon distinct from individual human beings) to humanistic studies of the process by which landscapes (and human beings) are socially created, within a context that views culture as a 'constitutive human process' (Williams 1980, 1982). Such studies require not only a rethinking of basic conceptual elements, but immersion in dialectical thinking.

The dance of the dialectic has caught geography within its flow in recent years, and we are now well rehearsed in its movements. The melody is polyphonic, voiced by self-professed humanists, pragmatists, realists, Marxists, structuralists, post-Marxists, post-structuralists and the like, but there is none the less harmony in the common quest to overcome Cartesian dualism, and to interpret the world through synthetic understanding. The concept of structuration, despite its flaws, has played a key role in bringing us to recognize the fundamental recursiveness of social conditions, and to transcend the simplistic dualisms that mark debates concerning 'structure and human agency' or 'individual and society' (e.g. Duncan 1985).

Few attempts have been made by geographers to apply a dialectical methodology to studies of landscape. The meaning of landscape is, in fact, to a remarkable degree taken for granted given its privileged position within our discipline. Geographical definitions of landscape abound, of course, but they are seldom critical, and an attempt to follow through upon the logic of these many definitions leads to many loose ends and contradictions.[1] The confusion is not an indictment of the profession, but rather reflects basic contradictions in philosophical approaches, as well as contradictions given by the term 'landscape' itself, which is variously interpreted as both idea and object, process and pattern, purpose and result. The challenge is to make landscape as all of these, as full dance production, but it may be necessary first to go back to the *barre*, to establish the foundations of the concept, and the possibilities for dialectical interpretation.

SOME PHILOSOPHICAL DIRECTIONS IN LANDSCAPE GEOGRAPHY

Despite the variety of definitions and philosophical perspectives represented, there seems to be general agreement that 'viewing' landscape is a selective process, that we 'must proceed in a sequence of "word packages" and can draw attention to only one aspect of the landscape view at a time' (Johnson & Pitzl 1981, p.214). A choice must be made. This choice is prescribed by epistemological limitations and ideological ordering.

The epistemological problem is rooted in the fact that both the limitations of human perception and the parameters of organized logic preclude comprehension at once of total reality, which is in a continuous state of becoming. Geographers must therefore address the fundamental problem that there is an apparent gap between reality and reason. Understanding does not grasp the whole, but requires some form of conceptual abstraction, or *analysis*, to be communicated. We therefore make analytical statements about a synthetic world. The concept of landscape is a geographical tool used analytically to organize material reality perceptually and conceptually, denoting the total environment, but expressing it in a partial way. This partiality is the basis for some very difficult philosophical issues, because it follows that analysis must proceed on the basis of choices concerning what is to be abstracted from the total. Such choices are informed by theories concerning how we account for the order that is imposed on the environment in creating landscape, theories based on fundamental assumptions concerning, *inter alia*, the existence of 'rules' according to which material transformation (formation of landscape) occurs, the role of human or other agents in bringing about such transformation, the relationship between consciousness and praxis in the enactment of that human role, and the ontological status of the respective contents of the material environment and the human mind.

Geographers addressing the above issues with regard to landscape have tended to frame their arguments in terms of either idealist or materialist philosophies. Idealist perspectives have dominated humanistic studies that explicitly recognize the efficacy of individual human action in defining relationship with landscape. Inspired by the statement that 'the most fascinating *terrae incognitae*

of all are those that lie within the minds and hearts of men'
(Wright 1947, p.15), a paper by Lowenthal (1961) pioneered the
expansion of landscape studies to include epistemic phenomena. It
was proposed that since 'the geography of the world is unified
only by human logic and optics' account should be taken of 'personal' worlds of experience, learning, imagination and memory,
the phenomena of which '(e)very image and idea about the world
is compounded' (Lowenthal 1961, p.260).

This insight instigated the turning away by geographers from
studies of holistic culture to studies of the qualities of existential
conditions. The past two decades have seen emphasis on
understanding human value, emotion and the quality of experience
in general, and have produced a new sensitivity to the human
condition. Unresolved in such studies, however, where they have
rested on subjective idealism, is the issue of how the partial world
of individual experience is connected to the realm of social,
political and material reality. To distinguish the subjective
'geography of the mind' from the world around is to assume a
distinction between the 'real' and the 'perceived', or between
nature and its mental image (Prince 1971b). Beneath this
assumption is a Cartesian separation of epistemic traits from
ontological status, or being, and a concommitant isolation of
action, experience and thought into distinct realms that are not
synthesized in accounting for the material production of both
landscape and social conditions. The resulting landscape as 'a
centre of meaning constructed by experience' (Tuan 1975, p.152)
is an abstraction, and idealist humanism is thus open to the
criticism that it fails to go beyond a critique of more mechanistic
methods of explanation to provide a comprehensive philosophy of
the world (Billinge 1977, Entrikin 1976, Gregory 1981).

Paralleling the development of humanistic approaches over the
past decade, Marxist approaches have responded to Harvey's
(1973) appeal to pursue a theoretical interpretation of the world
based on the exploitation of capitalism. Contained within that
appeal was a gem of an idea that should have set studies of
landscape at the heart of radical geography: 'One significant
element in this general process of differentiation is that *created space*
replaces *effective space* as the overriding principle of geographical
organization' (Harvey 1973, p.309).

It is unfortunate both that Harvey chose to emphasize abstract
space and thus set a generation of radical geographers on what
now seems to be a logical wild goose chase (since what is space
anyway but a nothing? cf. Samuels 1979); and that cultural
geographers did not see fit to recognize in Harvey's suggestion the

basis for a new interpretation of created landscapes. Instead, cultural geographers remained for the most part trapped in idealism, and Marxist geographers who would claim to redress the shortcomings of humanism have done so at the expense of rejecting individual action and experience as the constituent elements of social history, and of rejecting the landscape as an essential element in the production of social reality. Despite the emphasis in the works of Marx on a fundamental understanding of the material production of social conditions, Marxism in the structuralist form that has dominated geography (although by no means exclusively), has failed to recognize the importance of the quality of landscape, concentrating instead on the relations of production and social classes, presented in aggregate terms and determined by structures that originate outside human consciousness. For all that history depends, in structuralist Marxism, on the praxis of human beings to be produced, its terms are those of an external system that contains within itself the conditions of history (Duncan & Ley 1982, Eyles 1981). Rather than acting as a complement to the voluntarism of humanistic perspectives by releasing the potential to go beyond the individual in an incorporative way, therefore, Marxism has met humanism head on in a negation of the most appealing qualities of both philosophical systems.

As shown elsewhere in this book, and in a growing number of geographical publications, this standoff is being overcome, in what seems to be one of the most exciting conceptual breakthroughs in the discipline for some time. This breakthrough has come about in two major areas. The first is a focusing on the spatial/temporal aspects of social process most evident, despite its problems, in the rapidly expanding literature based on a theory of structuration. The second is in what Jackson (1986) refers to as the 'rediscovery of place' which includes local impacts of structural changes; experiences of places; the 'reciprocal link between society and space' (p. 121); as well as the political significance of human activity in place.

All these areas depend on a concept of landscape as the basic unit of analysis in a renewed regional geography. Re-examination of the landscape concept in such light requires an epistemology that allows for the incorporation of particular and general observations (Johnston 1984) and that transcends the philosophical problems referred to above. Furthermore, it requires both an explicit recognition of the fact that all concepts of landscape are ideologically laden and rooted in specific historical circumstances (Cosgrove 1983), and a methodology by which to understand the ways in which landscape is expressed as ideology. This chapter uses concepts put forward by Jean-Paul Sartre in an attempt to provide a heuristic basis for such a methodology.

THE INCORPORATIVE SOCIAL PHILOSOPHY OF JEAN-PAUL SARTRE

Two questions arise from the foregoing discussion. The first is, who sets the rules, human beings or some force or entity independent of human will? A broadly humanistic position asserts the anthropocentricity of human habitation of the earth, and the interdependence of the qualities of landscape and the qualities of human experience. A more specific position is required in specifying the terms of human habitation, and in answering the second question, which is how is it possible to mediate between the aggregate of all human action in organizing and experiencing landscape and the immediacy of human action as it actually occurs, as individual experience at a unique temporal/spatial moment? The aggregate is difficult to grasp because of its unspecifiable dimensions and thus is prone to reification; the immediate is impossible to grasp both because it is transcended in the contingent flow of experience, and because simple logistical problems preclude any but the most limited discussion of individuals.

The problem is an old philosophical issue. It is set out by Kant (1956, 1964, pp.48–55), who posits overcoming this contradiction between knowledge and object as the basis for truth, thus contributing to the ground rules for contemporary positive science. Understanding does not grasp the whole, but requires some form of conceptual abstraction, or *analysis*, to be communicated. For Kant, this process required that knowledge be categorized prior to its application to sensory (empirical) data, making clear the distinction between the form and the content of knowledge. Kant's legacy has been to set for science the task of making analytic statements about a synthetic world. The possibility of dialectical understanding (that is, where a distinction between the analytic and the synthetic is not only overcome, but also impossible) was for Kant merely the 'logic of illusion', a contemptible form of sophism.

As geographers address this dualism, perhaps a phenomenology of mind is less effective a methodology than an encounter with landscape. For is it not in the common organization of landscape that the mediation of the individual and society occurs? The terms of this organization are not self-evident, however, since it has no meaning until (a point which Lowenthal's proverbial hammer has so neatly hit on its head) it is unified through human action. Through logic and optics, we set out to achieve both an analytical selection of facts to be observed, and a synthetic integration of individual, society and material objects. This double-edged project provides the justification for Sartre's philosophical project.

Sartre dedicated much of his career to the project of overcoming idealism of all forms, whether as the assignment of a non-material status to knowledge or ideas, or the separation of knowledge from the act of knowing. Knowledge is depicted throughout his work only as praxis, engaged knowledge, a form (in fact the only form) of human relationship with the world. Knowledge and experience are synonymous. Sartre is most famous for his insistence on the freedom of knowledge, or consciousness, to define the quality of experience, as the projection of life toward the goal of life itself (Sartre 1956).

A second aspect of Sartre's work, one that is often overlooked or discounted, is his adamant claim that human existence occurs only through material praxis within the concrete material world. The fact that that world includes others creates the inescapable situation that human beings are social beings, and precludes the possibility that relationship with the world can ever be understood as any single consciousness in relation to any specific set of physical objects, except as an analytical abstraction. Sartre's major work, *Critique of dialectical reason*, is an unfinished attempt to overcome the duality of Cartesian reasoning, and the show the possibility, indeed the necessity, for a single, but dialectical, logic for thought and engagement with the world. The title of this work, derived from Kant's titles, is a statement of the impossibility of *either* pure or practical reason; there is only dialectical reason. The difference between internal (constituent) reason and external (constituted) reason occurs on a continuum, rather than as an abrupt switch of logical categories.

Within this logical framework, knowledge is always partial, necessarily contingent, purposive (if not purposeful), unbounded, and never separated from its concrete, material, historical circumstances. Knowledge and the world are not separate realities, but reality *is* the knowledge relationship. Knowledge is always *thought*, never *a* thought. (The problem with idealism of all forms, including that of logical positivism, lies in its attempt to create a status for knowledge independent of material context.) Furthermore, a combination of the necessity of plurality of consciousness with the necessity of dialectical transcendence leads to the rejection of the possibility of absolute truth occurring at any level of analysis. In this respect, Sartre is in complete agreement with Marx.

Within most Marxist interpretations, however, the quality of experience is determined not by the conscious appropriation of a necessary material world, but by the conditions imposed on consciousness by that world at the point that its physical properties become manifest in the process of production. This difference defines the critical point at which Sartre departs from historical materialism. It amounts to a rejection of a 'dialectic of nature', and denial of all forms of historical determinism, or of any intrinsic logic, such as the 'logic of capitalism', independent of human consciousness.

In thus refuting both idealism and materialism, the logic of Sartre's argument proceeds according to the principle of 'dialectical reason'. To render an extremely complex discussion in a few words, dialectical reason refers to what Sartre terms the 'intelligibility of praxis'. At the most fundamental level, human existence is also the intelligibility of praxis, in that all action is social, and constitutes a form of language, and is therefore at least potentially intellible. Because history is nothing but the totalization of praxis (occurring within a material world that is continually transcended by praxis) historical process can be understood as the condition of intelligibility of praxis. Praxis imposes on knowledge the synthetic engagement with the world (existence), as well as the possibility/necessity to transcend existence (becoming) through distantiation, in the mode of knowledge we refer to as *analysis*, by which particular moments of thought are *thought through*. Dialectical reason originates in individual consciousness, as the form of relationship between individual and environment, but it sets the conditions for all forms of social relationship (Sartre 1976, *passim*). This continuum proceeds from the interior of individual thought to the exterior of social structure. At each 'point' is imposed the need for *critical* interpretation of the world, and of ideas about the world.

The practical result of the principle of dialectical reason is that it allows a contextual interpretation of the world as relationship. This can occur at two scales, as a phenomenological description of human consciousness, or as an examination of history as social relations, but there is no absolute boundary between the two, and they are based on an *identical logic*. Dialectical reason thus encompasses the constituent terms of existence, that is, the possibility for human existence and its exigent conditions, as well as the constituted terms of history, where material conditions, and as we shall see structures, emerge. Put another way, the dialectic between individual and society, or between consciousness and the world, is expressed as a trajectory from 'the possible as infinite totality' (individual consciousness) to 'the possible as conditioned and limited field' (social history) (Chiodi 1976, p.41). Not only does this dialectic mediate the theoretical and the empirical, and the analytic and the synthetic, it also defines the contradictions of individual and society, past and future, regressive and progressive thought. This programme not only destroys the spurious association of humanism and historical materialism with the study of agency and structure respectively, but it allows an understanding of relationship that expresses not a duality but an interconnectivity. Far from lapsing into subjectivism or positing an individualist perspective (cf. Gregory 1982), this is a truly incorporative approach to social science. Furthermore, it is a critical perspective that allows exposure of the terms of reference of the analyst according to the same terms as those of society as a whole.

How do structures operate within this system? Sartre was a strong critic of structuralism, on the basis that it is non-dialectical. Relationships independent of consciousness may be necessary, but they must be dialectically related to consciousness. He criticizes Lévi-Strauss (1958) for applying his analysis only to systems that were already constituted, thus divorcing them from immediate consciousness of them and discrediting history. Foucault (1969), he claimed, failed to interpose praxis, which would mean to consider society as the passage from one structure to another, rather than simply to uncover successive structures.[2] Structuralist approaches are a reversion to a form of positivism wherein 'thought is only language, as if language itself were not spoken' (Sartre 1971, p.111).

If Sartre rejects structuralism for its inability to comprehend the synthetic totalization of history, he does not reject analysis of structures as a legitimate mode of historical investigation: an analytical mode in which the heuristic abstraction of structure makes understanding possible. Structures are real, to the extent that we experience a diachronic progression wherein structure comes about through the praxis of its agents, but then gains the power to condition subsequent praxis. Structure, simply put, consists of rules by which human actions are guided. We set the rules, change them, but also submit to them. They are materially real and exist as signification, through linguistic action, which includes both the production of formal language, and the production of landscapes, the physical setting for life. In being material manifestations, however, they are not inert. They are charged with meaning and heavy with ideology; their power lies in the knowledge of the consequences should they be broken, and their power is wielded in exact proportion to the cumulative praxis invested in them, and to the extent that that praxis was itself structured.

Sartre also claimed that 'superstructures' constitute regions of relative autonomy above the relations of production and praxis. Their autonomy derives not from an external dialectic, however, but from the fact that human beings have created them and, through praxis, given them the power of relative autonomy. This is how the condition of alienation occurs. Structures thus construed can be interpreted on two levels, first as:

> the moment of structure, where totality *appears* as the thing without man (sic), a network of oppositions in which each element defines itself by another one where there is no term, but only relations of difference (Sartre 1971, p.111).

But analysis of structure as a momentary manifestation must be followed by a second level of analysis where:

> . . .it can no longer be a question of structures already there which will exist without us. . .Structures impose themselves on us only to the extent that they are *made* by others. Thus, to understand how a structure is made, it is necessary to introduce praxis as that totalizing process. Structural analysis must pass over to a dialectical comprehension (Sartre 1971, p.111, emphasis added).

A dialectic of nature is then refuted:

> In nature you will not find oppositions such as those described by the linguist. Nature knows only the independence of forces. Material elements are placed one next to the other and act one over the other. But the lines of force are always external (Sartre 1971, 111).

The human system, on the other hand, is one that originates in internal relationship, and is sustained only through the active engagement (exteriorization) of human will. If we accept that all engagement of the environment (action) is simultaneously a linguistic statement, then language becomes the fundamental mode of socialization. As social, structured beings, we are never independent of our history and our history only functions as it is linguistically acknowledged in each moment of praxis.

The importance of analysing structure comes in knowing the role that it plays at the moment that it maintains order in history, for a structure's efficacy occurs in direct relation to its ability to maintain order. A structure is non-existent if unacknowledged, even if acknowledgement consists of denial or misunderstanding. A structure must be *known* although this prerequisite does not give it the status of pure thought. In being known, or incorporated within the sphere of human action, it gains historical presence; however, at that very moment, 'history is already on the way to undoing it' (Sartre 1971, p.112). To put it another way, human beings are somehow developed by the development of structure, 'in all moments totalizer and totalized' (Sartre 1971, p.115).

The real problem, then, is not in simply uncovering (knowing) structures, but in going beyond structures to understand what Sartre terms 'depassement', the perpetual integration and reintegration by which structures are created, transcended, and given conditioning power, and by which their consequences become subject to rational enquiry. To stop at structures would be to create an 'intermission' in the realization of the depassement. That is what has occurred in the modern Marxist project, and that, Sartre claims, is a 'logical scandal' (Sartre 1971, p.115). Sartre's *Search for a method* (1963) was an attempt to rectify this scandal.

An important concept in this dialectical method is that of 'process'. A process is an abstraction, an analytical device, by which synthetic, totalized history is organized for the purpose of analysis, and structural conditions are uncovered. It is the assignment of pseudo-organic character to a particular range of moments whose existence cannot, of course, be made absolute linguistically, but can be depicted. This is why we must deny absolute truth, but affirm concrete understanding. In everyday life, we unreflectively engage in process creation continually. For example, we claim to 'go for a walk' and in doing so acknowledge and communicate the pseudo-organic character of the walking process. Linguistically, we turn the walk into a thing, without thereby reducing the walk to each of the steps required in completing the circuit, or each of the muscle contractions required to produce the walk, and so on. At the same time, we engage an entire historical context in which going for a walk holds a specific set of structured meanings. We walk to obtain exercise, to ponder a problem, to vent frustrations, and so on. We can therefore understand walking as a process, but only by understanding the context for a particular walk can we uncover the degree and manner in which *that* walk is structured.

The designation of process in this way is more than a simple statement of the constraints of language. It illustrates the condition of life as a constant dialectical interplay between the analytic and the synthetic, where, in projecting ourselves between past and future, we also move between the totality that threatens to engulf, and the organization that makes rationality possible. In so doing, we cast ourselves into the 'flow' of experience, where language, being and action are synonymous, adelphous, but not undifferentiated, for they are organized through the mechanisms of process and structure. The constitution of process (like structure) is therefore a necessary aspect of engaging the world, the means by which we recognize, give meaning to, our actions.

The researcher utilizes process in a similar, but more critical, manner to denote, or make into a thing, that which is analysed. We speak of the 'process of change', the 'process of moderniza-tion', or the 'process of urbanization', and we recognize that these are abstractions. None the less, they are legitimate abstractions if we wish to convey history as pseudo-organic phases in order to provide a basis for communication. We cannot convey totality, but we can convey a pseudo-totality through the concept of process. This does not change the fundamental material character of all language as praxis (or praxis as language), but it does illuminate the dilemma we face that time and space, as the qualitative conditions of material development, cannot themselves be made objects. The best we can achieve is then a homogeneity of comprehension and development, but it is important in doing so not to confuse analytical organization with empirical totality, in either a practical or a critical sense. (Or to believe that by reducing it to its *pliez* we capture the essence of the dance.)

In depicting structures in this way, Sartre provides a logical justification to overcome not only dualism, wherein structures are viewed as either determinate of or determined by human agency, but also duality, wherein an ameliorative, or recursive, relationship is posited between structure and human agency. The latter position reflects Giddens' 'structuration' concept (Giddens 1976, 1977, 1979, 1981, 1984) which has provided a catalyst for conceptual breakthroughs in geography (e.g. Gregory 1981, 1984, Pred 1983, 1984a, b, Duncan 1986), yet has been very limited in its empirical applications, and has gone only part way to solving the duality issue (Gregson 1986, S. Smith 1984).

The 'duality' of agency and structure is mis-specified if the relationship between the two is seen as merely recursive, each constituting the other, for such a depiction provides a basis only for understanding the constituted terms of contingent history, not the constituent terms according to which history is possible. Whereas Giddens, therefore, provides a framework for analysis of emerging social conditions, he does not provide the basis for understanding how human beings possess the potential to be part of social conditions. He begins with a theoretical, rather than a philosophical perspective (Gregory 1984, pp.119–20). Sartre's work allows us to connect the two, by creating a link between the possible and the empirical. This link gives access to history and geography, as well as the spatial/temporal organizations of the social.

TOWARD A REDEFINITION OF LANDSCAPE

Geographers concern themselves with a certain range of structures that define the quality of the human being/environment relationship. One of the major concepts that allows us to do this is landscape. The methodological logic outlined above can be applied to the concept of landscape to achieve two tasks. The first is the conceptual clarification, and by corollary philosophical justification, of the term within the geographical lexicon. The second is the confirmation of landscape as one of the central methodological entrées to an understanding of human habitation of place.

To achieve the first purpose, landscape can be understood in exactly the same way as the concept of process just described. It is the assignment of pseudo-organic quality to the ongoing dialectical relationship among a set of material elements. Landscape is a process. Its quality is *pseudo*-organic, because unlike simple nature, something that is historically transcended cannot also be organic in itself. It carries with it an irreducible context. It is therefore a totalization at any historical moment. Its qualities are limited, however, by the power of human perception, and landscape is not therefore 'everything', but a choice of inter-connected elements within a totality. Its status is analytic. Landscape is a process organized, as Lowenthal noted, logically and optimally, while at the same time it is always a concrete, material transformation of nature.

We can then speak of the 'urban landscape', the 'agricultural landscape', 'landscapes of despair', or 'fear', or the 'physical landscape', as particular organizations of the environment by geographers. In doing so, it is not necessary to idealize landscape as so many generations of cultural geographers have done. It is quite possible to avoid idealism simply by specifying that we refer to moments of organization that are immediately transcended, but that none the less can be expressed linguistically. At the same time, it is not possible to reduce landscape to its constituent elements, or to a set of structures, since the dialectical quality of existence makes such a reduction a logical impossibility (a 'logical scandal').

What is important, however, is to achieve an analysis of landscape as homogeneity: of comprehension and development, praxis and process, form and meaning, structure and human agency. To achieve homogeneity, or integration, is to uncover rationality, on which analytical understanding is based, and from which synthetic understanding can proceed. Rationality can be defined after Sartre's discussion in *Critique of dialectical reason* (1976), as a form of congruence between understanding and language. To make something rational is to reveal its structures, as they are expressed in its physical manifestation. The way in which landscape is most useful to the geographer is as a form of language that structures and is structured by the production of social relations. This applies not only to the geographer, but to all those who inhabit the landscape. Habitation becomes expression.

An aspect of landscape as language that is of especial importance to geographers is its relative durability. We need to recognize not only the theoretical principle that life is a continuous process of landscape formation (a necessary but insufficient starting point), but also that the process according to which landscape emerges includes the very forceful imposition of physical conditions that structure the actions of human beings over long periods of time. In this sense, a historical understanding of landscape formation is necessary to link the qualities of particular landscapes to both their social meanings and the technological means by which they are controlled (or not controlled) at particular times and places.

The creation of landscape comes about through human labour in its broadest sense, the process by which human beings produce and reproduce their lives. Labour expresses the dialectical relationship with the natural environment:

> its possibility and its permanent necessity rest upon the relation of interiority which unites the organism with the environment and upon the deep contradiction between the inorganic and organic orders, both of which are present in everyone. Its primary movement and its essential character are defined by a twofold contradictory transformation: the unity of the project endows the practical field with a quasi-synthetic unity, and the crucial moment of labour is that in which the organism makes itself inert (the man applies his weight to the lever, etc.) in order to transform the surrounding inertia (Sartre 1976, p.90).

Thus materiality is the bond, forged by labour, by which the relationship between the individual and nature is constituted. The relationship is reciprocal, a vast symbiosis, wherein humans are '"mediated" by things to the same extent as things are "mediated"' by them (Sartre 1976, p.79). The relationship, however, is always both social and univocal. That is, people share the landscape they create through their labour, and only humans are capable of initiating the act of totalization by which the environment is organized in a purposeful manner. They have projects. In the absence of human labour, the world would be a meaningless and mechanical collection of energy.

The concept used by Sartre to indicate the surrounding world of material possibility (the raw material of landscape) is the 'practico-inert':

> . . .every object, in so far as it exists within a given economic, technical and social complex, will in its turn become exigency through the mode and relations of production, and give rise to other exigencies in other objects (Sartre 1976, p.189).

This is *the world*, created by and embodying (in a crystallized form) human praxis in fulfilment of its projects, a world of objects that surpass their objective quality because they possess a history, and therefore human significance. They have been imbued with the power to act upon both human beings and other objects through praxis.

Every human action is constituted as 'primarily an instrumental-isation of material reality' (Sartre 1976, p.161). It is a totalization undertaken with specific tools and technology and, as such, a statement of an entire ensemble of social relations. The statement occurs in the landscape as signification:

> . . .precisely because signification takes on the character of materiality, it enters into relation with the entire Universe. This means that infinitely many unforeseeable relations are established, through the mediation of social practice, between the matter which absorbs *praxis* and other materialised significations (Sartre 1976, p.161).

In nearly all situations, the exigency of the surrounding landscape is diffuse and complex. Sartre gives the example of deforestation in China (1976, pp.162–4). The Chinese peasants removed the forest cover over large sections of the country in the positive project of producing greater agricultural yields. In so doing, they humanized their landscape, inscribing it with the significance of their actions, embodying a cultural tradition. At the same time, they created the unforeseeable effect of massive flooding, and the peasants' positive actions became the means of their own destruction, which nothing in their set of cultural structures could circumvent. This set of events was possible only within the historical material ensemble that produced a society in which certain tools, certain methods of labour and certain projects occurred as the cumulative result of preterit action that was also more (in terms of its consequences) than it was intended to be. Human action was united in the material environment, inverted and turned into an exigency that set new conditions and alienated the peasants from the land that was their means of survival. In this sense, the landscape is capable of usurping the human attributes of those who sustain it and give it power.

It is in understanding the diffusion of power within the landscape that geographers face their greatest methodological challenge. In discussing method, I wish to focus on only two aspects of the linguistic expression of landscape, as crystallized praxis and as ideology. To term landscape 'crystallized praxis' is nothing more or less than to recognize that it is the totalized appearance of all praxis at a particular moment. It has a history, or as Samuels (1979) calls it, a biography. Since the depassement of its biography occurs only as actual physical manifestation, it is open to investigation. It can be read, or interpreted. There are practical constraints to interpretation, but these do not prevent legibility in principle.

We may therefore recognize that:

> Given the fact that. . .landscapes are typically the product of design. . .i.e. the result of many decisions by many people over long periods of time, it is very striking indeed that they 'add up'. They develop coherence, they are instantaneously recognizable; if one knows the cues, a single slide can be enough to identify a place (Rapoport 1984, p.56).

This intelligibility is an intrinsic aspect of all praxis as human beings engage their surrounding materiality, organizing it to produce meaning. The organized landscape 'adds up' in the literal sense that it is composed of *every* praxis that has been initiated toward it, nothing more. Every praxis, however, has meaning, draws upon former meanings, and transcends its own moment. Through this process, structure is engaged to maintain order.

The clarity of the landscape message, however, is problematic.

The landscape is fraught with contradiction and new contradictions are created each moment, and interposed at three levels. First, human consciousness may be defined as the synthesis of contradiction, not in the Marxist sense of an imposed alienation, but in the sense that it involves a mediation of internal and external reality. Consciousness becomes itself as the constant mediation of this contradiction. Second, as the above example shows, in the production of the material object, praxis is alienated from itself. Its power is usurped, made inert by the material world, thence to be returned in a transformed state. The world acts back on the univocal relationship established by human praxis. Third and most importantly, praxis refers not to the single act of an individual, but to the totality of all those who inhabit landscape. Individual intentions are seldom expressed intact, but are countered by the actions of others and by the unintended results of former actions.

In the interpretation of landscapes, therefore, geographers are faced with the task of trying to ascertain, among other things, the levels of ambiguity according to which it was created, the degree of contradiction that is manifest, and the efficacy of the linguistic statement in putting forth intelligible messages. All of these qualities rest on the structure of landscape, as the rules by which actions are guided. Efficacy of structure depends on its clarity, and the amount of power that it wields. Clarity of structure, as in any form of language, is best achieved when a high level of formality exists. The Japanese formal garden, for example, is created according to a set of rules that is well established in historical tradition, and relatively inflexible in form despite infinite capacity for expression. These rules have been created in such a way that their strength lies in their aesthetic appeal. The Japanese garden is thus a poetic form of language, and as such plays a powerful but limited cultural role.

If the power of structure is measured by the degree to which it is commonly understood and expressed within the landscape, then it is ideology, as the form of structuring and sharing value, that underlies the power, or effectiveness, of landscape as a linguistic statement. Landscape as ideology takes as many forms as human beings are able to construct in exerting power over others. A totalitarian landscape, for example, is an expression of alienation, where ideology is made to be an autonomous superstructure with relative control over the agency of social individuals. We may envision another form of landscape in which ideology is expressed as common project, shared concern, the fusion of praxis to create congruence of structure and the agency of individuals. Geographers have identified particular landscapes that emerge when an organized ideology is imposed to create specific political goals as in 'Mao's China' (Salter 1976), or where particular groups effect a common ideology at the local level in opposition to the imposition of institutionalized structure (Kobayashi 1983, Salter 1981). Historical processes such as 'post-industrialism' may provide the basis for structuring of landscape as a way of life (Ley 1980). In each case the landscape is the depassement of ideology, a statement of the social relations of its production and the ground for further development of geographical landscape studies.

The geographical challenge of modernity is to recognize ways in which particular landscapes emerge, and are organized, as expression of the conditions of the world at this time. Technological advances in communication, for example, allow dominant ideological systems to gain hegemonic influence over landscapes in far-flung corners of the world (Santos 1983). At the same time, never before have human beings been better equipped to use global communication as a means of addressing common concerns. Technological organization, with all that it implies, is therefore one of the most important aspects of contemporary landscape analysis, and must be taken into account in addressing the environmental expressions of contemporary world issues, some of which have been addressed in this volume, such as hunger, the restructuring of the labour force and the livability of cities; all issues which will influence the character of future landscapes.

A second aspect of the geographical challenge of modernity lies in the fact that all linguistic statements, including landscape, are ideologically laden. Lefebvre has pointed out with reference to 'space' that it is 'shaped and molded from historical and natural elements. . .It is a product literally filled with ideologies' (Lefebvre 1976, p. 341). Furthermore, as Cosgrove (1985) has shown, those who analyse landscape, no matter what their *form* of discourse, are drawn into its ideological implications. As inhabitors of landscape we are subject to the norms that it imposes, and as analysers of landscape we necessarily make normative judgements. A critical approach to landscape analysis, based on an interrogation of its means of discourse, is thus the only means by which we, as geographers and human beings, can gain control, through understanding, of the power that language exerts.

CONCLUSION

Sartre's argument is summarized in the following statement: 'What is essential is not that (human beings) are (made), but that (they make) that which made (them)' (Sartre 1971, p.115). The making of landscape as shared, organized human activity provides one of the fundamental bases for geographical study. The landscape itself provides material manifestation of the intelligibility of human action, and is a means by which historical conditions can be interpreted and understood. Sartre's work can be used by geographers to further this project, in three ways: (1) ontologically, by providing a set of heuristic concepts for a dialectical understanding of the developing world as relationship among conscious human beings and the total ensemble of material objects; (2) epistemologically, by providing a rationale by which to establish intelligibility as the basis for understanding; (3) methodologically, by providing the analytical concepts of process and structure, which can be expanded upon when investigating spatial and place-specific components of social organization at a number of scales.

Without advocating intellectual reductionism, it can be suggested that Sartre's work be extended and made more relevant to the needs of social scientists through the incorporation of more specific theories of social action (such as, for example, that of Giddens), and social discourse (as found in various ways in the linguistic work of Barthes, Foucault and Baudrillard). These writers are by no means entirely compatible, and it is beyond the scope of this chapter to show their interconnections. What is important is that they share(d) with Sartre a commitment to understanding human development in terms of. its possible – Sartre's words, constituent – terms, as well as its empirical – constituted – history, and according to the dominant ideological systems by which its conditions are structured. The geographical theory of landscape can provide the third component in a triad of action, discourse and object, for a comprehensive, dialectical understanding of social history.

This *pas de trois* is far from a full stage production. For geographers, certainly, the dialogue has only begun to allow the choreography of events that make up the process of landscape. Having said this, I am reminded that the dancer who remains too long at the *barre* never develops the presence that is required of the dance in its fullest sense. Necessary as it may be to continue honing our conceptual categories, it is also necessary to go out into the landscape and continue efforts at empirical understanding, to engage in dialogue according to the terms that landscape is made. Then may we *see* in landscape the qualities of human creativity.

NOTES

1 See, for example, Appleton (1975a, b), Cornish (1935), Duncan (1980), Duncan & Duncan (1984), Gregory (1978), Hartshorne (1939), Jackson (1964, 1980), Johnson & Pitzl (1981), Lowenthal (1978), Lowenthal & Bowden (1975), Lowenthal & Prince (1965), Meinig (1976, 1979), Salter & Lloyd (1977), Tuan (1977), Wright (1966). Recent critical examinations of the place of landscape studies in geography include Cosgrove (1984), Ley (1981, 1983a, 1985a), Meinig (1983), Relph (1981), Samuels (1979).
2 Sartre is usually depicted as being opposed to both structuralists and post-structuralists (see, for example, Aron 1975, Hirsh 1981, Kurzweil 1980). Such claims are perhaps relevant if confined to Sartre's earlier work, *Being and nothingness* (1956), but it has been clearly shown that the debate between existentialism on the one hand (e.g. Warnock 1965, Ch. 6, pp. 135–81) and Marxism on the other (Jay 1973) is not appropriate for Sartre's work (Chiodi 1978, Meszaros 1979). Sartre must be seen in opposition to all critical theorists – such as, for example, Habermas – who see meaning and intelligibility as centred anywhere but in human consciousness. None the less, recently Mark Poster (1984, pp.27–8) has suggested that Foucault and Sartre were coming closer to one other, and that perhaps their differences have been highlighted according to the interests of other theorists who have used one or the other to grind their own axes.

SECTION III
Directions

Introduction

This final section provides a retrospective on the intellectual paths that have led to the present historical moment in geography, as well as some prospective thoughts about the discipline. Dangerous as such projects are, and fraught with possibilities both for mis-specification of trends or of facile interpretations of extremely complex directions, these chapters manage both to 'round off' the present volume, and to set a major part of the agenda for re-making the future.

The educative function of a reflexive, theoretically and contextually informed geography is also central to the first chapter in this section, by Denis Cosgrove. Cosgrove states that his objective is to examine 'the historical links between European humanism and geography since the Renaissance, and to do so from a perspective that is broadly informed by the categories of historical materialism'. He argues that 'humanism and the humanities are neither as individualistic nor as universal in their values as they often are proclaimed to be, but rather that they have to be understood within the historical contexts in which they have emerged and in terms of the social role that they have played as emblems and instruments of ideology and power, deployed within the historical struggle between competing social groups'. Cosgrove directly challenges the practice of humanists and humanist geographers, stating:

> humanism has always been very closely aligned in practice with the exercise of power both over other human beings and over the natural world. In large measure geography has developed as one of the instruments of that power. At the same time . . . both humanism and geography offer the opportunity to subvert existing power relations. This potential derives from the role that both have always played and continue to play in education. . .

The final two chapters take up the question that dominates the second section, which is how we generalize. Andrew Sayer and

David Ley complement each other in showing that although we have come a long way toward convergence, we have many issues left to address, and significant points of divergence. Both chapters emphasize that attempts to do historical materialism and humanism have required a rigorous examination of method, as well as the development of analytical concepts that transcend the rigid categories established in narrower and earlier approaches. The ground for debate is to a large extent laid out within these two chapters.

10

Historical considerations on humanism, historical materialism and geography

DENIS COSGROVE

INTRODUCTION

Any humanist endeavour is inevitably historical and, to a degree, reflexive for it concerns the nature and purposes of conscious human subjects together with their individual and collective biographies. Whether such an endeavour is also critical is a matter of theoretical predisposition, although I would argue that a truly reflexive study must inevitably be critical, for its duty is to leave no assumption or aspect of human conduct unexamined. If this is so, then the claim made by increasing numbers of geographers in recent years for a geographical humanism, or a humanistic geography (not necessarily the same thing), impels some degree of historical understanding on their part of humanism and the humanities on the one hand, and of geography on the other. It also requires an examination of our motives for adopting the particular epistemological and philosophical positions to be associated with humanism. Curiously, while discussion of epistemology has been considerable and detailed over the past decade, discussions by geographers of the history of Western humanism and its relations to geography seem to have been relatively few in number and sketchy in execution. Of these studies, two of the most thoughtful have come from Edward Relph (1981) and David Ley & Marwyn Samuels (1978). However, none of these writers has a strong historical orientation in his research and, more significantly for my concerns in this chapter, none has sympathy for, or uses concepts available from, the most critical of the available historiographic traditions: historical materialism. In earlier papers I have sketched the lines along which I believe a critical encounter between humanism and historical materialism within geography might move (Cosgrove 1978, 1983). My intention here is to examine in greater detail the historical links between European humanism and geography since the Renaissance, and to do so from a perspective that is broadly informed by the categories of historical materialism. Above all I am concerned to show that humanism and the humanities are neither as individualistic nor as universal in their values as they often are proclaimed to be, but rather that they have to be understood within the historical contexts in which they have emerged and in terms of the social role that they have played as emblems and instruments of ideology and power, deployed within the historical struggle between competing social groups. As a humanity, geography too has been implicated in this process of social production of knowledge and its use within social struggle.

The period which I will examine in detail is the 15th- and 16th-century Renaissance in Italy and England, because the cultural movement that we call the Renaissance was characterized above all by a revived and self-conscious humanism; a celebration of man (sic) as the centre of the created cosmos and the elevation of the human body, mind and spirit to a place in the order of things hitherto reserved for God alone. The Renaissance marks the historical origin of modern Western humanisms. It was equally a critical period for geographical development, both practical, in the forms of exploration and discovery, surveying and mapping, landscape change and design; and theoretical, in terms of new and refined concepts of environmental and spatial relations. From the explanatory perspective of historical materialism such vast and significant alterations in human practice and conception cannot remain ungrounded in the realities of social production and power relations, and it is to these that I relate Renaissance humanism and Renaissance geography (Cosgrove 1985). And this is not merely of antiquarian or academic interest. It is of direct contemporary relevance because humanism has always been very closely aligned in practice with the exercise of power both over other human beings and over the natural world. In large measure geography has developed as one of the instruments of that power. At the same time, and herein lies the subtlety of the dialectic, both humanism and geography offer the opportunity to subvert existing power relations.

This potential derives from the role that both have always

played and continue to play in education. A humanist education incorporating the study of geography as the understanding of the richness and variety of the human world is individually and socially liberating. Precisely for this reason such an education has, in most European societies, been reserved for the ruling class and its progeny. The liberating potential of humanism and geography is too dangerous to the interests of ruling groups to be made freely available to all. It has consequently been constantly subverted in the interests of sustaining power. This argument I will develop below. It leads us inevitably to the conclusion that in order to realize the democratic and liberating possibilities of both humanism and geography, humanists and geographers must equally confront the intellectual and practical implications of historical materialism. In doing so they will be no mere playthings of an already self-sufficient social theory. Rather, humanists and geographers themselves have the potential to broaden and refine the existing theory and methods of historical materialism.

RENAISSANCE HUMANISM AND RENAISSANCE SOCIETY

What a piece of work is a man, how noble in reason, how infinite in faculties, in form and motion how express and admirable in action, how like an angel in apprehension, how like a God! the beauty of the world, the paragon of animals (*Hamlet*, II,2).

In the opening years of the 17th century, when Shakespeare expressed so lyrically a pivotal belief of the Renaissance: that man stands at the very centre of creation, the humanist philosophy he celebrated was more than a century old. To some extent it was by now under attack from both sides of an increasingly intolerant religious divide. Renaissance humanism was far more complex than is often recognized today, with aspects which may surprise those accustomed to associate humanism and the humanities with individual sensitivity or ivory-towered self-indulgence. Today scholars of Renaissance humanism recognize that many of the classical texts which were so revered by the self-styled revivers of antiquity were well known and widely read during the Middle Ages. These included such key works as Vitruvius, Ptolemy and the Corpus Hermeticum as well as Plato, Aristotle and Pythagoras and the Greek and Latin moralists, dramatists, poets and historians. Renaissance scholars also tend today to discount the idea of an intellectual 'great leap forward' in 15th-century Florence and find the origins of the Italian Renaissance as far back as the 12th century and even earlier. Nevertheless, what remains significant about the humanists of Florence and, later, other Italian cities like Ferrara, Padua, Mantua and Venice in the 15th century is their very self-consciousness in proclaiming a new age of learning, their determined search to recover and translate from the earliest sources the works of the ancients, and the leadership in this movement of lay people (Martines 1980). All these aspects indicate a widening of access to knowledge, a widening enormously stimulated by printed books after mid-century. They also indicate a significant change in ruling class culture as learning – or at least the proximity to and patronage of learning – came to compete with physical prowess, equine and military skills in the self-image of rulers and their courts. The Renaissance was always its own best publicist, which is not to say that we must take it entirely on its own terms. Rather we should recognize the conscious attempt to promote a new age with a distinct culture of humanism, and interrogate that age and culture with the intention of uncovering the foundations of its consciousness in the order of 15th- and 16th-century European societies.

The most significant social fact of late-medieval Italian communes in which the new learning developed was their commitment to trade and the accumulation of mobile capital. To be sure, capital was invested in land and urban real estate as well as long-distance trade, but even land investment was seen in terms of accumulation and rates of return as much as in terms of feudal dues and land as status and disposable income. The social formation of the Italian communes was, as Marx himself recognized, characterized by merchant capitalism in the cities and transitional, semi–capitalist relations on the land. The formal structures of feudalism had long since loosened in Upper Italy. In the fluid conditons of the Italian communes from the 12th to the early 15th centuries, social mobility for the merchant and artisan was greater than at any time before or after, at least until Italian Unification. The existence of a significant bourgeoisie and its growing self-awareness accounts for certain aspects of Italian urban culture in the 14th century, notably the *Stilnovo*, the new vernacular style in poetry associated with Dante Alighieri, and *Il vero*, the new realism in painting of which Giotto was the finest exemplar (Baxandall 1972). But the humanism of the 15th century was far more closely aligned to the courts of *signori* and princes like Cosimo de' Medici in Florence than to merchant houses and artisan guilds. Such courts were the product of the takeover of formerly independent communes by individual family clans, *signori*, who exploited the inherent instability of earlier urban democracy to seize absolute power. But for all their imposition of individual authority the rulers of the Renaissance city states were unable to unravel three centuries of social and economic history. They could not call upon a disciplined, seigneurialized countryside to reimpose feudal obligations and forms of exploitation, for land was owned in large measure by urban bourgeois and cultivated by tenants or sharecroppers. Aristocratic feudal values carried no legitimacy in the city (Anderson 1974). Thus the chivalric and militaristic cultural activities and the ideology appropriate to the royal and aristocratic courts of contemporary France, England or the Empire were inappropriate in the Italian cities. Court culture had to compromise with the social realities of a merchant and artisan citizenry.

The Renaissance *Zeitgeist*, of which humanism was an important part, may be understood as the outcome of this cultural compromise. Humanism drew upon the lifeworld and self-image of the urban bourgeoisie and interpreted the world in ways that made sense to people of that class. The popularity of Plutarch's *Vitae*, biographies of great men of action first published in 1470, is but one example of how the results of classical philology could speak directly to the quattrocento Florentine or Bolognese (Giustiniani 1985). At the same time humanism was aligned to, sustained and legitimated, the realities of autocratic and oligarchic power. As Italy witnessed decelerating economic advance and failed to complete the transition to capitalism during the 16th century, we observe humanism adopting an increasingly aristocratic and exclusive flavour. This social context of humanism is perfectly summed up by Martines:

> Humanism was the city-state's major intellectual experience. It was a refinement of values that derived largely from the urban ruling groups, but it also needed their confirmation to have any force as a program. Humanism spoke for and to the dominant social groups. It was a concerted study of classical Roman and Greek literature, it embraced the whole field of knowledge from poetry and geography to natural science, but its syllabus was heavily weighted in favour of 'the humanities' and its philosophical focus was man in society (Martines 1980, p.191).

The humanities and the philosophy of humanism were, in modern parlance, relevant. Educated Italians believed that the secret of civil government was to be found in their country's history, in Classical Rome and its intellectual mentor, Greece.

Therefore, the solutions to contemporary problems of society in the Italian cities were to be read in the texts that survived the barbarian destruction of the ancient world. Similarly, the concept of *virtù*, so central to Alberti and Machiavelli, celebrates the victory of individual will, its capacity to determine human destiny. Its classical exemplars taught lessons to the modern prince. This was no antiquarian quest, for the results of humanist research bore directly on the realities of power and its exercise. Humanists consequently found their employment in the courts; advising rulers, counselling and educating the younger scions of ruling families, writing laudatory histories and deploying their skills of rhetoric in the public presentation of rulers. Machiavelli may have become a popular synonym for devious and immoral statesmanship, but in his devotion to the survival of the Prince he typifies the *raison d'être* of the humanist.

The central conception of any humanism is of course its conception of man. It is a truism to say that the Renaissance placed man at the centre of the cosmos, made man 'the measure of all things'. But there are few better phrases to capture the essence of the difference between a medieval world picture centred upon God wherein men and women held an important but decidedly inferior position in the great chain of being and in which their prime duty was to transcend the mortal world and their mortal bodies, and the new image of man as the central miracle of creation, the perfect measure of God and God's creation, whose duty was, by exercise of reason – the supreme faculty – to know God through knowing and perfecting his creation. The distinguishing feature of the human individual was the capacity to reason, and human reason through the exercise of the arts revealed the consistency and pattern written by God in creation, a pattern that reflected the nature of the divine being himself. The most convincing intellectual reasons for this humanist philosophy were to be found in classical authority and mathematics, and I shall return to them below. For the moment it is important to recognize the critical links between this new way of seeing the world and the place of humans within it and the social formation of the Italian city states. Humanism challenged directly the notion that status derived from birth alone, the notion that social rank represented a pre-ordained order, that it was the secular reflection of the divine hierarchy and its temporal expression in the ecclesiastical hierarchy of Pope, cardinals, archbishops, bishops, priests and people. In this respect humanism was a socially progressive force because it emphasized the common features of all humans (at least that half which was male); their capacity to reason and to apply reason to civil life. Humanism may be seen as the characteristic self-conception of an urban bourgeoisie, of 'new men', standing in contrast to the typical social conceptions of feudalism.

At the same time, in practice, humanism was from the outset aligned to the exercise of power. Humanists found their employment in the courts of the princes. Cosimo de' Medici, the banker's son turned prince of Florence, was the paradigm employer of humanists, sending them across the trade routes to bring back ancient texts, using them as his closet counsellors and specifying the pattern of philological study to be followed by one of the greatest of the 15th-century humanist scholars, Marsilio Ficino. And, naturally, given their dependent position, the humanists underwrote the interests of their employers, acting in their speeches and writings as the ideologists of signorial power. Leon Battista Alberti is typical of this as he argues that it is reason and *virtù* that make for a full development of humanity, characteristics potentially to be found in every person, but going on to point out that only some are truly gifted in this area, and that for such people fortune also seems often to smile in her distribution of worldly wealth and power. The message of humanism is, then, ambiguous, its egalitarian strain apparently too subversive to be allowed full rein by those who hold power. As we shall see, this feature is by no means confined to the Renaissance period.

HUMANISM AND THE HUMANITIES, HISTORICAL AND APPLIED

Thus far I have used the term humanism fairly loosely, making no distinction between humanism as a way of seeing the world and the particular activities of those who studied the humanities, the *humanisti*. While not wishing to pursue in detail the enormously complex set of issues thus raised (for a full discussion see Giustiniani 1985), it is necessary to distinguish between Renaissance humanism in the broadest sense as equivalent to the *Zeitgeist* of the Italian Renaissance, and the study of the humanities as the revival of ancient learning for its own sake, as practised by the teachers of classical philology: the *Humaniae Letterae*. The latter, developing first in Italian and later European universities, was resolutely academic. It 'aimed neither at transcendence nor practical purposes' and thus generated no philosophy of man or society, although it was clearly informed by the same spirit of enquiry into the past as the broader humanism. However tightly defined and separate the academic humanities were, they fed the broader humanism in three important ways: in their concern with history they nourished a conception of time as progressive and of the present as a rebirth toward a better future; in making available correct and authenticated texts and providing commentaries on them they offered models of conduct to the present; and in linking study of the humanities to education they offered the possibility of applying those models and ensuring the progressive advance of society. The task of mediating between the work of these classical scholars, or humanists properly called, and the vibrant urban world of bourgeois, mercantile and commercial Italy was filled by what we refer to as humanism in the broader sense. If we define the former humanism as historical we may speak of the latter as applied, despite the somewhat anachronistic flavour of the word.

Applied humanism was far closer to a philosophy of man than was historical humanism. Drawing upon the medieval conception of an earth-centred globe and its circumrevolving terrestrial, celestial and supercelestial spheres, the humanists mathematized it by reference to Platonic, Neoplatonic and Pythagorean concepts of number and geometry. The process was initiated in the writings of Nicholas of Cusa (Cassirer 1964), taken up by L.B. Alberti, stimulated enormously by Marsilio Ficino's translations of Plato and the Corpus Hermeticum, and most elegantly articulated by Pico della Mirandola in his *Oration on the dignity of man* (1487). Here was truly a Renaissance philosophy of man – a humanism promoted and sustained not so much in the universities as in the academies, those new centres of intellectual discourse that originated with the Platonic Academy in late 15th-century Florence and which had spread to all the major Italian cities by the end of the 16th century (Yates 1983). In this philosophy man had a definite place in the order of creation, partaking of all its elements and free to choose his fate. The province of learning in the Renaissance academies was wide, for the conception of the world was holistic. A key concept was the unity of all things via correspondence, harmony and proportion. A consistent pattern of creation was to be found most clearly expressed in the theorems and axioms of Euclid's *Elements*. In its more speculative aspects this theory tended toward transcendence: the worlds of the astrologer, the magician, the alchemist, the conjuror and other occultists who feature so strongly in the world of the late Renaissance (Yates 1979). But equally the interests of the academicians were practical. The same principles of number, the same geometrical relations which fired their esoteric discourse, were also applied to human interventions in the natural world, to shaping the world as the home of man. They informed the painter in the use of perspective; the merchant in the calculation of volumes and weights, profits and losses; the banker in determining interest rates; the navigator in using his astrolabes and charts; the cartographer in fixing co-ordinates; the engineer in measuring stresses or designing pulleys and beams, raising water or designing machines; the artilleryman in fixing his trajectories; the surveyor in levelling and calculating land shape and area. Above all, as the Roman writer Vitruvius had shown, number and proportion were the foundation of architecture, the 'queen of the arts', the one whereby mankind emulated the creative act by making its own microcosms – the human temples built in the image of greater temples; the human body and the earth itself (Barbieri 1983). It is within this applied humanism that we discover geography as a humanist science.

GEOGRAPHY AND HUMANISM

Geographie teacheth wayes, by which, in sundry formes, (as Sphaerike, Plaine or other), the Situation of Cities, Townes, Villages, Fortes, Castells, Mountaines, Woods, Hauens, Rivers, Crekes, and such other things, upon the outface of the earthly Globe (either in the whole, or in some principal member and portion therof contayned) may be described and designed, in commensurations Analogicall to Nature and veritie: and most aptly to our vew, may be represented. Of this Arte how great pleasure, and how manifolde commodities do come vnto vs, daily and hourely: of most men, is perceived. While, some, to beautify their Halls, Parlers, Chambers, Galeries, Studies, or Libraries with: other some, for thinges past, as battels fought, earthquakes, heavenly fyrings, and such occurentes, in histories mentioned: therby lively, as it were, to vewe the place, the region adioyning, the distance from vs: and such other circumstances. Some other, presently to vewe the large dominion of the Turke: the wide Empire of the Moschouite: and the little morsell of ground, where Christendome (by profession) is certainly knowen. Little, I say, in respecte of the reste &c. Some, either for their own iourneyes directing into farre landes: or to understand other mens trauailes. To conclude, some, for one purpose: and some, for an other, liketh, loueth, getteth, and vseth, Mappes, Chartes, & Geographical Globes. Of whose vse, to speake sufficiently, would require a booke peculier (Dee 1570, n.p.).

This description of the nature and uses of geography is taken from the *Mathematicall Praeface* by the Elizabethan Renaissance scholar John Dee. The *Praeface* was published in 1570 to accompany the first translation into English of Euclid's *Elements* and is intended to indicate to English readers the breadth of application of this classical text on geometry. Dee aims his discussion precisely at 'unlatined students' without experience of the universities, the practitioners of his day: the mechanics, instrument makers, navigators and others who may now access a major theoretical text in the vernacular. A similar task had been performed by Nicolo Tartaglia in 1543 when he translated and introduced Euclid to a vernacular Italian readership.

The description of geography opens that part of the *Praeface* where Dee lists the 'mathematical arts' dependent upon a knowledge of Euclid. This is essentially a classification of the applied sciences (see Fig. 10.1), the nature and value of each of which is summarized. As a mathematical art geography 'by Mathematicall demonstrative Method, in Numbers, or Magnitudes, ordreth and confirmeth his doctrine, as much and as perfectly, as the matter subject will admit'. Geography is eminently a practical art, locating places and events occurring on the earth's surface, globally or regionally, and representing them at an accurate scale ('commensurations analogicall to nature and veritie') according to geometrical principles of projection. There is a distinct echo of the political and military dimensions of this work, particularly in the reference to the relative sizes of the Turkish, Russian and Christian parts of the world. Dee was very much involved in Elizabethan imperialism, promoting English navigations across the Atlantic and toward China via the North-East passage (Taylor 1930). In this he exemplifies the dual reference of humanism, promoting the widespread dissemination of practical knowledge and allying it with the promotion of national power and prestige. The frontispiece to Dee's *General and rare memorials pertayning to the perfect arte of navigation* (1577) shows Queen Elizabeth steering the ship of Europe toward Empire under the auguries of beneficent constellations.

What is most significant about this passage is its location in a document that very clearly summarizes – for the first time in English – the humanist philosophy of the Renaissance. In the opening lines of the *Praeface* Dee gives voice to the place of mathematics in the humanist cosmos: 'A meruaylcus newtralitie have these thinges MATHEMATICAL and also a straunge participation betwene thinges supernaturall, immortall, intellectual, simple and indivisible: and thynges naturall, mortall, sensible, compounded and divisible' (Dee 1570, n.p.).

Dee's acknowledgements to Pico, Ficino, Agrippa and many other Italian humanists make his acceptance of this humanist philosophy beyond doubt and we know from his biography how committed he was to this holistic way of seeing the world and our place within it (French 1972). In locating geography within the humanist scheme of knowledge Dee reflects the position ascribed to it by Italian authors. But we should not take the practical definition of geography too restrictively for, as we have seen, applied Renaissance humanism was above all a holistic philosophy. Later in the *Praeface* Dee refers to geography uniting with astronomy to yield cosmography, the understanding of the whole of creation. From this it is but a step to an environmentalism grounded in analogical reasoning in which man as microcosm is united with the macrocosm of natural world and cosmos, animated throughout by the divine breath. It was this mode of thinking which found expression not only in the practical arts but also in transcendental practices and utopian hopes for a universal religion to bind the schisms of Counter-Reformation Europe (Yates 1972). Renaissance humanism stands at the turning point of the modern world. It was the base above which the scientific revolution was to be erected (Debus 1978). While Renaissance humanists believed in the simultaneous possibility of human harmony with the course of nature and human control over the world, it was the latter alone that triumphed in the unity of science and technology, while the former was suppressed, intellectually by Enlightenment rationalism, and practically in the interests of private appropriation through the market.

HUMANISM AND EDUCATION

Dee's *Praeface* was intended primarily as an educational document, promoting the new knowledge, both practical and speculative, and celebrating the vernacular as the linguistic means of reaching the widest audience. Dee's disdain for the universities was considerable and reflected that of many of the philosophical humanists (Yates 1964). The classical humanists in the universities remained bound to the ancient languages, their knowledge restricted to a specialized readership. It was the broader humanism of the academies that sought a greater public participation in

Here haue you (according to my promisse) the Groundplat of
my MATHEMATICALL Praeface: annexed to *Euclide* (now first)
published in our Englishe toungue. An. 1570. Febr. 3.

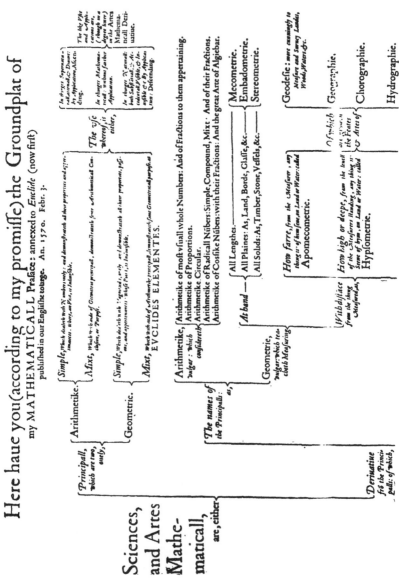

Sciences, and Artes Mathematicall, are, either

Principall, which are two, onely,

- Arithmetike.
  - Simple, Which dealeth with Numbers onely: and demonstrateth all their properties and appurtenances: where, an Vnit, is Indiuisible.
  - Mixt, Which with aide of Geometricall principall, demonstrateth some Arithmeticall Conclusion, or Purpose.
- Geometrie.
  - Simple, Which dealeth with Magnitudes, onely: and demonstrateth all their properties, passions, and appurtenances: Whose Point, is Indiuisible.
  - Mixt, Which with aide of Arithmetike principall, demonstrateth some Geometricall purpose: as EVCLIDES ELEMENTES.

The vse whereof, is either,

- In thynges Supernaturall, eternall, & Diuine: by Application, Ascending.
- In thynges Mathematicall: without farther Application.
- In thynges Naturall: both Substãtiall, & Accidentall, visible, & inuisible. &c. By Application: Descending.

The like Vse and Applications are, (though in a degree lower) in the Artes Mathematicall Deriuatiue.

Deriuatiue fró the Principalls: of which,

The names of the Principalls: as,

- Arithmetike, vulgar: which considereth
  - Arithmetike of most vsuall whole Numbers: And of Fractions to them apperteyning.
  - Arithmetike of Proportions.
  - Arithmetike Circular.
  - Arithmetike of Radicall Nübers: Simple, Compound, Mixt: And of their Fractions.
  - Arithmetike of Cossike Nübers: with their Fractions: And the great Arte of Algiebar.
- Geometrie, vulgar: which teacheth Measuring
  - At hand
    - All Lengthes. —— Mecometrie.
    - All Plaines: As, Land, Borde, Glasse, &c. —— Embadometrie.
    - All Solids: As, Timber, Stone, Vessels, &c. —— Stereometrie.
  - With distance from the thing Measured, as,
    - How farre, from the Measurer, any thing is: of him sene, on Land or Water: called Apomecometrie.
    - How high or depe, from the leuell of the Measurers standing, any thing is: Seen of hym, on Land or Water: called Hypsometrie.
  - Of which are growen the Feates & Artes of
    - Geodesie: more cunningly to Measure and Suruey Landes, Woods, Waters. &c.
    - Geographie.
    - Chorographie.
    - Hydrographie.

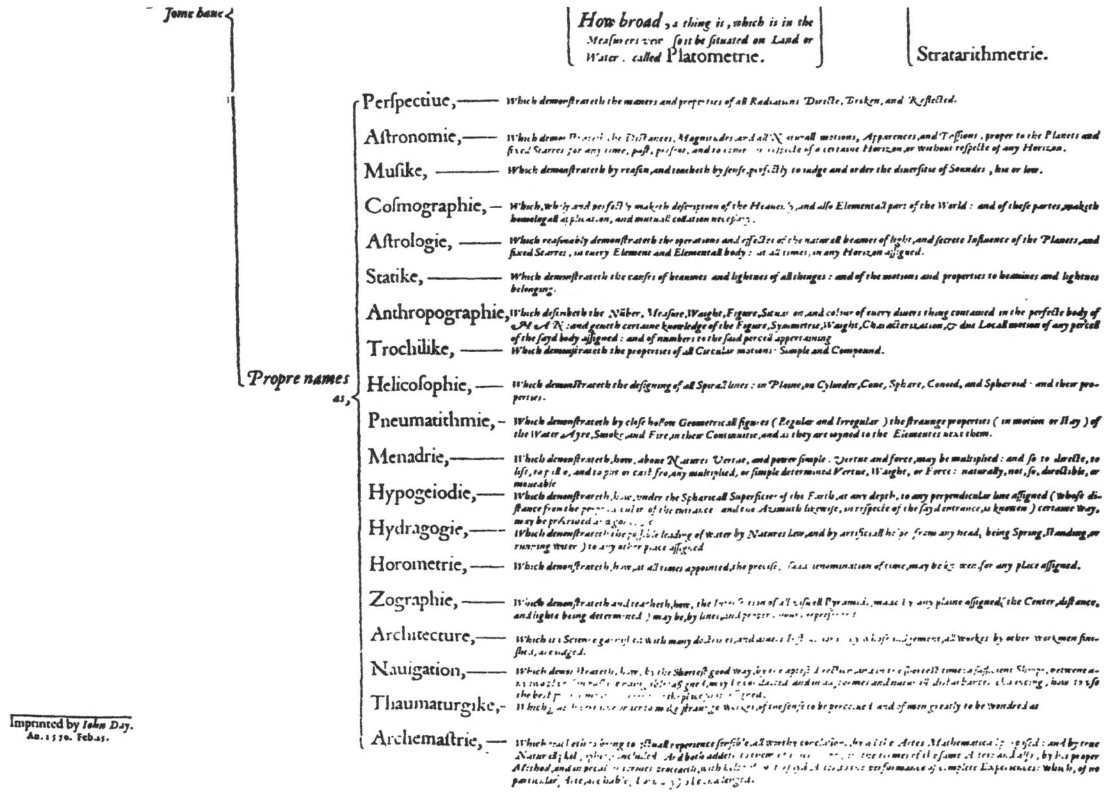

Some base {

Propre names as,

How broad &c. a thing is, which is in the
Measure... (as be situated on Land or
Water ... called Placometrie. }

Stratarithmetrie.

Perspective, ——

Astronomie, ——

Musike, ——

Cosmographie, ——

Astrologie, ——

Statike, ——

Anthropographie,

Trochilike, ——

Helicosophie, ——

Pneumatithmie, —

Menadrie, —

Hypogeiodie, —

Hydragogie, —

Horometrie, ——

Zographie, —

Architecture, —

Navigation, —

Thaumaturgike, —

Archemastrie, —

Figure 10.1

knowledge and learning than had hitherto been possible. One of the most consistent interests of the Italian and later the French academies was the refinement and consistency of national vernacular language – again an example of the ambiguity of humanism in its simultaneous links both with political power – in this case providing one of the key mechanisms of state control and the dissemination of a national culture – and with progressive human liberation – in this case promoting wide secular access to knowledge through publishing it in the words of everyday speech.

From the beginning the Renaissance *Umanisti* had been concerned with education. They translated Greek educational theorists like Isocrates and Plutarch, and reflected on the pedagogical ideas of Socrates and Aristotle. New schools and new methods spread from Italy into France, Germany and England (Giustiniani 1985). In the 19th century under the influence of Goethe, Schiller and Winckelmann *Neuhumanismus* generated a new revolution in pedagogy based on a revived conception of classical and Renaissance humanism, and in our own century Rudolf Steiner drew upon the ideas of the Renaissance academies to develop an educational philosophy for the 20th century. As these examples show, the relationship between humanism and education is historically consistent. Throughout, a humanist education has been regarded as fulfilling the Greek ideal of *Paideia*, the realization of the whole human potential, intellectual, spiritual and physical. It is essentially a secular, but not materialist, conception and one that is inherently liberating and egalitarian in its treatment of all humans as potentially capable of discovering and enlarging their humanity through the disinterested pursuit of learning.

Equally consistent historically is the ambiguity of humanism between human liberation and human control. I have stressed its initial connections with the practice of political power, with the control of some groups over others. This is to be found in the writing of the earliest Renaissance humanists. In Ficino's *Platonic theology on the immortality of souls* (1482) the writer celebrates human power not only over nature and animals but over other humans:

> Single animals scarcely suffice for the care of themselves or briefly of their offspring. But man so abounds in perfection that he rules himself first, which no beasts do, then governs his family, administers the state, rules peoples and commands the entire world. And as though born for ruling he is entirely

impatient of servitude (quoted in Martines 1980, p.216).

In the celebration of *virtù* Renaissance humanism could scarcely make this link more explicit. Inevitably this connection has borne on the educational practices of humanism. From early on, and despite the populism of men like Dee, the radical potential of a humanist education has consistently rendered it unsuitable for the broad mass of the population, at least in the eyes of those with power. The Italian academies, initiated with a membership drawn from a wide spectrum of status groups, became increasingly aristocratic over the course of the 16th century (Ventura 1964). Yet an education in the humanities has remained, since the 16th century, the favoured preparation of the ruling classes of Europe and to some extent America, while a pedagogy for the mass, where it has been available, has always tended to insist on the passive learning of moral codes and technical skills, training its subjects for a predetermined slot in the economic and social order. *Liceo classico* in Italy, the *Ecole des Hautes Etudes* in France, the study of Greats at Oxford, such have been and remain the favoured routes into the most powerful and influential positions in their respective societies. In contemporary Britain the Thatcher counter-revolution in its impact on education has borne acutely on the teaching of the humanities in schools, polytechnics and universities, suppressing them while celebrating and promoting the value of technical and business training above all else. Yet it has left classical and humanist studies unscathed in the public schools and ancient universities where they will still provide the educational experience of the nation's ruling elite. A humanist education does promote that self-confidence and fitness for rule that Ficino refers to. No wonder the humanities loomed large in Arnold's programme for Rugby in the early years of the last century, the programme that through its adoption by the English public school system became the training for the imperial ruling class of England. The example could be endlessly repeated.

GEOGRAPHY, THE HUMANITIES AND CONTEMPORARY EDUCATION

Most discussions of the nature and philosophy of geography operate on the unstated assumption that they concern a discipline whose principal activity is scientific research (e.g. Johnston 1983). Even discussions of humanist geography assume that the problems of epistemology and methodology they address concern geography as a research discipline. The point has been made already in geography that true humanism does not yield clear methodologies in the way that positivist views of science may be able to (Harris 1978). Yet whatever conclusions we reach about geographical research, our discipline remains overwhelmingly a pedagogical concern. It became instituted in the universities of Europe and North America in the last years of the 19th century, principally for the training of teachers. This occurred at a time when the traditionally unified academic pursuit of humanist wisdom was giving place, particularly in the new civic universities, to separately conceived disciplines valued for their training of a skilled managerial and technical class. The very notion of geography as a fundamental research discipline dates from the postwar period and the particular social context in which higher education and the social sciences found themselves (Ackerman 1945, Harvey 1974, 1984). In the furious theoretical debates of the past 15 years geographers seem collectively to have forgotten the centrality of their role as teachers in the purpose and practice of their discipline. This is less excusable in the case of humanist geographers than in that of any other group.

The principal tasks which confront humanist geography now are these: to define the role that geography should play in an education seeking to promote human self-knowledge and reflection on our place in society and in the natural scheme, an education committed thus to realizing the human liberation always implicit in humanism; to remain aware and critical of the historical and contextual pressures to which geography like any other division of knowledge is always subject; to struggle for the widest social availability of humanist education, and for the eradication of elitism and exclusivity in both content and curriculum. These tasks inevitably demand a confrontation with historical materialism because as a social theory historical materialism is committed to the same ends of human liberation, offers the most coherent and historically consistent understanding of the contextual conditions in which geography as a discipline and the objects of geographical examination operate, and is consistent in its rejection of hierarchy and exclusivity. Above all historical materialism acts as a control on those tendencies within all humanisms toward solipsism, the uncritical proclamation of universal human values ungrounded in history and hermetically sealed against criticism, toward moralism and toward utopianism.

By the same token, in confrontation with historical materialism, humanism upholds such fundamental areas of discourse as morality, aesthetics and human passion against the tendency within Marxism toward a crude materialism and economism. In geography as in other areas of study we have seen how easy it is for historical materialism to congeal into a theoretical system, a self-proclaimed predictive science subject to the very criticisms so often levelled against bourgeois thinking. 'A humanistic approach has to refute scholasticism as something which exhausts itself in the elaboration of rational schemes and paradigms that not only fail to explain or unravel reality but also have absolutely nothing to do with it' (Garin 1969, p.9). However, there is nothing to be gained by seeking some theoretical resolution to the divergent tendencies within humanism and historical materialism; such a resolution would be purely rhetorical. The tension between these two ways, one of knowing and one of explaining, can be a creative one, in research to be sure, but above all in teaching our geography as a celebration of the diversity and richness of the natural world and of the places that human societies create in their attempts to make over that world into the homes of humankind.

11

On the dialogue between humanism and historical materialism in geography

ANDREW SAYER

INTRODUCTION

In this response, I want to limit myself to the main theme of the book, the dialogue between humanism and historical materialism. I am sure many of us have grown weary of the standard tripartite division of social scientific or geographical methodology into positivist, interpretive/humanistic and radical positions, and hence such a dialogue is long overdue. I shall argue first that a dialogue is only possible between certain varieties of humanism and historical materialism, and that the insights of critical theory are particularly useful in identifying these variants. Another issue relevant to humanism and historical materialism which has been perceived most acutely by critical theory is the divorce of science from the realm of practice and values. In the second section I shall argue that humanism and historical materialism as represented in geography tend to reinforce rather than reverse this divorce. Finally I shall discuss some more mundane methodological aspects of the problem of synthesis, and of the attempt to interrelate the concerns of humanistic and Marxist research. However, in approaching these issues, I find myself starting from a different position from many of the other contributors to this volume, including those who, like me, share an interest in realist philosophy. It may therefore help if I explain some of the differences.

First, I find the commitment to geography somewhat strange,

particularly in view of the enthusiasm of many of the contributors for 'synthesis'. I know it is sometimes difficult to disentangle professional interests from academic issues but the combination of the goal of synthesis with the reproduction of a single discipline produces a strange beast, one which derives from extravagant and imperialistic claims for the scope of the discipline. Disciplinary boundaries and jealousies merely get in the way of enquiry: why reinforce them?

Second, and related to this, having been interested for many years in the relationship between historical materialism and interpretive philosophies of human action and social science, and having observed equivalent debates in geography, I have been depressed by the reluctance of many geographers – both Marxist and humanistic – to read non-geographers and learn from philosophers, sociologists *et al.* In part, this chapter is a commentary on the consequences of that isolationism.

My third point concerns realist philosophy, which is drawn upon and criticized by several contributors to this volume. I confess to being slightly puzzled by these and other early reactions to realism, although of course books are always read in ways which differ from those which their authors intended (no bad thing in itself). Here I am referring not just to responses to my own *Method in social science* (Sayer 1984), but to the work of more authoritative writers such as Bhaskar (1975, 1979, 1986) and Keat & Urry (1975). Despite modern realism's sympathetic relationship to interpretive or humanistic philosophies of social science, humanistic geographers seem to ignore realism. Perhaps less surprisingly, Marxists tend to ignore the many concessions and accommodations which it makes to these anti-naturalist philosophies (even in Bhaskar's *The possibility of naturalism*) and concentrate on details of method, in the case of my own work, seizing on the distinction between internal and external relations in particular. All readings are selective but these interpretations of realism are selective in ways which ignore its insights on the problem of the relationship between humanism and historical materialism.

WHAT KINDS OF HUMANISM AND HISTORICAL MATERIALISM?

The most surprising absence is critical theory, although humanistic Marxists with affinities to it such as Sartre and Raymond Williams are cited by several authors, particularly Audrey Kobayashi. It is surprising because critical theory's principal concerns have included the relationship between hermeneutics and historical materialism, the divorce in contemporary society of science from ethics, and the associated attenuation of reason and dehumanization of science. It therefore helps us understand the reasons for the loss of what Denis Cosgrove calls the 'humanist endeavour' in so much of contemporary social thought. In what follows I shall draw extensively on arguments from critical theory.

Arguments for a convergence between humanistic or interpretive approaches and historical materialism have been around for a long time – at least 20 years. What is so depressing about the situation is that convergence has been obstructed by problems, many of which have long ago been resolved in the literature on the philosophy of the social sciences (Benton 1977, Bernstein 1976, Fay 1975, Gellner 1973, Giddens 1976, 1981, Goldmann 1969, Gregory 1978, Habermas 1971, Hughes 1980, Keat & Urry 1975, Mattick 1986, Sayer 1979a, 1984, Ch. 1, Smart 1976, Williams 1977).

On the Marxist side, one of the main obstacles has been the inscrutability of dialectics or at least the apparent inability of its advocates to provide an extended and careful explication of what it involves. Few of its advocates pay attention to the many different concepts of dialectic implicit in the literature.[1] Often it functions as little more than a slogan, as in Harvey's tendentious references to 'the tough rigour of dialectical theorizing' (Harvey 1987). Sometimes it is presented as a way of overcoming dualisms, as in Kobayashi (this volume). Even if we agree that all the dualisms are bad – and this tends to be assumed rather than argued – it is not shown how they can be overcome in a way which does not conflate that which is actually distinct. Matters are no better on the humanistic side, for as Daniels (1985) notes, with a few exceptions (such as the work of Cosgrove, Pickles, and Jackson & Smith)[2], much of what passes for philosophical argument amounts to little more than assertion and name dropping. Here again, ignorance of recent philosophy of social science has resulted in considerable wastage of effort.

Not surprisingly, the scope for dialogue between humanism and historical materialism depends on what kinds of humanism or Marxism are involved. Some kinds of humanism exclude causal explanation and refuse to evaluate beliefs or practices, while others permit such combinations of understanding, explanation and criticism. Similarly, historical materialism embraces a considerable variety of positions, some of them, such as 'scientific socialism' antagonistic to humanism, some, like humanistic Marxism, sympathetic toward it. Such terminology is potentially misleading, so let me begin by saying that in criticizing 'scientific socialism' I am not objecting to the need for rigour and for theory, but to the way in which this need is seen as being met. In any case 'science' is a notoriously tendentious term, beloved of those who want to avoid providing a reasoned defence of their methods. Nor do I see humanistic Marxism as 'soft', atheoretical or unrigorous. However, whatever the limitations of the terminology, I wish to argue that dialogue is only possible between critical and humanistic variants of humanism and Marxism respectively.

Such an argument depends on the critique of interpretive social science which, as noted earlier, is now well developed in the philosophy of social science literature. Rather than drag through this critique yet again, I shall briefly list some of the more important conclusions:

1 The social is not reducible to the individual. Even individual subjectivity is structured through intersubjective relations and expressed in socially available concepts and language. Consciousness never consists of presuppositionless understanding, nor can it be reduced to expressions of private beliefs; for it always has structure or syntax, whether linguistic, conceptual or artistic, and such structures have a certain autonomy.

2 Humanistic or interpretive approaches to social science cannot avoid causal explanation, even though their primary aim may be the interpretation of meaning. Causal explanation is not inimical or irrelevant to the understanding of meaningful or meaning-dependent phenomena in society. To know what is going on in a seminar one must understand the concept of seminar, though this does not explain why one was held. Causally explaining why a particular seminar was held presupposes this understanding and in no way undermines the account of the meaning of 'seminar'.[3] Reasons have to be understood, but they can be causal. To deny this is contradictory because the act of reasoning in this denial presupposes the possibility of causing people to change their minds.

3 We cannot ignore the way in which structures and processes not alluded to in actors' accounts and actions may nevertheless be necessary conditions (and unintended consequences) of their actions.

4 In interpreting sets of ideas which are inconsistent or at variance with actual conditions, errors or illusions cannot be ignored without failing to show why actions produce the results they do. In other words, the attempt to restore or recover meaning inevitably slides into the reduction of illusion, and hence into

critical social science. This need not imply either a denial or a misunderstanding of actors' beliefs; criticism need not imply misunderstanding. If it did, there would be no point in humanistic geographers criticizing positivists and Marxists.

Although these points have been widely rehearsed, I believe more needs to be said about the historical materialist side.

The first problem here arises from a widespread failure to grasp the intrinsically meaningful or concept-dependent character of human action. Thus, to use a hackneyed example, the act of voting presupposes an understanding of what voting means and what elections are. So Marxism has to be interpretive as well as explanatory. This isn't a question of reserving a special place for 'ideology' or 'culture' while continuing to ignore the matter elsewhere, for even the most basic of economic relationships (e.g. exchange) presupposes that actors possess some understanding of the meaning of the action, though of course this need not be sophisticated. This point has again been argued forcefully for years, and yet studiously ignored for just as long by many.

This problem is related to a tendency to reduce human activity to work or struggle, neglecting communicative interaction, the sharing of experience and the negotiation or imposition of meanings and identities. Marxists have always rightly attacked those who would reduce social life to communication or 'symbolic interaction', as if societies did not also have to be reproduced through work, but Marxists often fall into the converse error of reducing communication (in the broad sense of sharing meaning, whether verbally or non-verbally, intentionally or unintentionally) to work. On this view labour 'becomes a paradigm for the generation of all categories' (Habermas 1971): childcare, education, literature and other 'cultural productions' have all been treated in this way, but none is wholly reducible to a labour process.[4] And what use are production categories for understanding phenomena such as the geography of fear, black consciousness movements, urban symbolism or urban riots?

As Habermas argues, the rationality underpinning science and its application in work is instrumental in character. The particular concepts used have to be understood, and the manipulations or experiments must be successful, but the former stand in an external relation to the latter – merely as means to an end. As long as the ends are achieved, that is all that matters.

The nature of this relationship can be seen more clearly when compared with that between knowledge and its object in the interpretation of meaning. Here the subject–object relationship is more like a dialogue. Unlike instrumental knowledge, the goal of the encounter is not primarily or immediately one of control, but of understanding another person or society through recovery and negotiation of meaning. At the limit, no social scientific or lay knowledge can avoid this; even if someone wants to manipulate you, they have to try to make sure that you understand their words and actions in the way they want you to. So communication in the wide sense, including social research, presupposes a degree of mutual recognition if it is to be successful (Habermas 1971, 1979, Winch 1960). If we are not prepared to grant those we study the respect of at least provisionally accepting anything of what they say as what they *mean* to say, communication and understanding cannot even begin. Here the relations between means and ends is not external and arbitrary but internal. Our own understandings are both means and ends. Yet given our habitual way of thinking of 'understanding' and 'knowledge' as merely an external description or reflection of the social world, rather than as something constitutive of it, the internal relation may not be readily apparent.

Consider, however, an example, that of describing someone by reference to a certain role, e.g. 'housewife'. This is more than a matter of applying an external label; it can actually shape their lives and objective possibilities. Roles are not only defined in terms of objective access to material things, like the means of production, but are also established through intersubjective meanings. A certain set of concepts are constitutive of what being a housewife is in the sense that they underpin the rules which a person follows in playing that role. Moreover, what you are depends not just on what you have, together with how you conceive of yourself, but on how others relate to. you, on what they understand you to be and themselves to be (e.g. 'husband' or 'breadwinner' in this case). Roles and social relations are rarely unilaterally established. So the concept 'housewife' is not simply an external description but a social *identity*. Again, it is one thing to argue that the material is primary, but absurd to deny or ignore the way in which social relations are structured by (and in turn structure) intersubjective meanings. Marxism has to recognize this primacy without falling, in the name of 'science', into a crass materialism which overlooks and suppresses, both in theory and practice, all the other interactively established dimensions.

Failure to acknowledge the intrinsically meaningful character of human practice supports the tendency to see revolution as a tool which changes society from outside by force (which it must be in part) but makes it difficult to recognize how it involves self-change. 'External', objectifying theory helps us control and change things and people as objects. But a theory which recognizes the intrinsically meaningful character of human behaviour helps people change according to their grasp of the theory and the way in which this affects their self-understanding. The difference is summed up in the contrast between a Marxist group which sees its role in seizing power and a women's consciousness-raising group.

Now although 'scientific socialism' is generally thought of as 'more materialist' than humanistic Marxism, the divorce of meanings in society from practices and material conditions in the former actually generates a kind of idealism. Meanings both confirm and are confirmed by material practices and conditions. Conceptual distinctions arise and are confirmed or modified through practice, or rather through our differentiation of a host of practices. The success of the distinction between private and public property depends on what happens when we try to distinguish them in practice, though equally, making the distinction in practice requires some understanding of the conceptual distinction. Nothing in this argument undermines the fundamental point that ideas are different from and not to be confused with matter; but where human practice is concerned they are nevertheless interdependent. Sometimes it may be quite harmless to separate out the 'level' of meaning from the 'level' of practice, but it can easily render opaque the relationship between thought and practice.

It follows from this that struggles against oppression are not only about material control (access to the means of production, etc), although this is invariably fundamental; they are also about identity and hence involve a *contestation of meaning* about what it is to be black, Islamic, or whatever. This is also why disputes over sexist language are far more than semantic matters; they influence how we relate to one another and hence what kinds of people we can be.

So meaning and identity cannot be 'etherealized' as a relatively autonomous level of ideology, in which actions, including struggle, are reduced wholly to physical behaviour and conflict. The 'sensuous human activity' which Marx argued Feuerbach's materialism could not grasp does not or should not refer just to the self-conscious activity of labour, but to communication in its broadest sense. As Raymond Williams once put it, the problem with 'vulgar materialism' and 'scientific socialism' is not that they are too materialist, but they are not materialist enough. In tearing meaning from its medium of material subjects and objects – real people in specific situations and social relations – they echo the bourgeois alienation of our knowledge from ourselves and our portrayal as objects, rather than subjects, of history.

These arguments concerning the concept-dependent nature of human action differ from many humanistic accounts in an important respect; they do not assume that meanings and identities in society are freely negotiated by actors. Even where they are not imposed or contested but accepted (perhaps actively accepted) this does not mean that they can be taken to represent individuals' freely chosen interests; to a considerable extent people have to adopt meanings, roles and identities which pre-exist them.

This viewpoint has implications for the way in which *power* is conceptualized. The recognition and explanation of power has traditionally been regarded as a strength of Marxism and a weakness of humanism, and rightly so. Yet there are certain kinds of power which tend to elude the categorical framework of scientific socialism. Although Marxism is extremely effective in identifying what Giddens terms 'allocative' power, which centres around the control over key material resources, particularly the means of production, it often overlooks 'authoritative power' which centres on constitutive meanings relating to roles, such as that of priest, teacher or patriarch (Giddens 1979). In many, perhaps most, cases, the two kinds reinforce one another; thus when threatened, patriarchal authoritative power turns on coercion. Conversely, allocative power is legitimized by appealing to concepts of authority. But scientific Marxism's ignorance of constitutive meanings forces it either to dismiss oppressive forms of mutual definition based on authoritative power or to reduce them to the allocative variety.

Earlier we argued that in scientific socialism the relationship between theory and society was taken to be external, ignoring conceptual, internal relations between 'scientific' concepts and the intrinsically meaningful practices of everyday life. This has two serious consequences. First, it supports a tendency – strongest in Althusserian variants – to put science on a pedestal and derogate and dismiss everyday knowledge and experience. It is one thing to argue that we need to be critical of and push beyond the unexamined assumptions and concepts of everyday thinking, but it is quite another to call for a 'science' which is radically distinct from everyday social knowledge, on a different 'terrain', as Althusserians used to say. Unawareness of the interpretive character of subject–object relations in social research dissociates theory from practice and supports a politically dangerous dismissal of everyday thought and practice. As Sheila Rowbotham and others have argued, these problems are paralleled by a tendency, sometimes evident in the personal behaviour of revolutionaries, to reduce communication to transmission on the assumption that there is nothing to learn from those less 'scientifically conscious' than themselves (Rowbotham *et al.* 1979).

The second consequence is closely related to this and derives from the assumption that there are only external relations between social theory and its object and hence between the means and ends of change. Historical change is seen entirely in terms of struggle over allocative power, in terms of control over objects. Winning this struggle requires internal discipline and external force, and a grasp of a powerful theory – one which is essentially instrumental in the sense used by Habermas. In other words, the implicit notion of change and revolution in the scientific socialist view takes the revolutionaries, and perhaps the masses, to be tools, agents. What is important in this view is not that they change themselves but that they gain, through practice, the 'scientific knowledge' to work out the 'correct line' and create the organization to seize power.

On a humanistic view of Marxism, self-change is more central. As a democratic form of association is an end or goal of struggle, it should not be struggled for in a way which negates that goal, as scientific socialist groups are sometimes wont to do. The means must be internally related to the ends by prefiguring them; the intention is to start as we mean to go on. Not even the most humanist of Marxists, however, believes that socialism can be achieved without considerable and decidedly undemocratic reaction from ruling interests; liberation must therefore still involve a struggle for power, though not at the price of suppressing the very goal that it is directed at attaining. Power struggles do require discipline and unity. Hence there is always an uneasy balance in any radical movement between achieving this at the same time as encouraging openness to criticism and democratic control.

Thus in humanistic Marxism, the process of historical change is not seen purely in terms of class struggle over the control of the means of production, or indeed, other material resources, but in addition, as a process of rational argument, on a social scale, involving a widespread contestation of meaning and identity, of concepts of authority, status and gender, of the nature and scope of politics and indeed of democracy itself. Once again, all these issues are intimately bound up with material structures, resources and practices. For example, the scope for extending democratic control into new spheres, such as the organization of production, is limited by their material and technical character. Nevertheless, the questions of meaning and identity remain as part of the picture and the central role of rational argument in identifying constraints and illusions counters the tendency of struggles based solely on the seizure of power to become corrupted by that very struggle. This is not at all to suggest that argument and critique can replace struggle; as Benjamin said '. . .the revolutionary struggle is not fought between capitalism and mind. It is fought between capitalism and the proletariat'. Still, humanistic Marxism rightly warns us about the importance of keeping the development of what that process should be *for* in the foreground.

THE DIVORCE OF SCIENCE FROM PRACTICE AND VALUES

According to Denis Cosgrove, the humanist endeavour 'concerns the nature and purposes of conscious human subjects. . .'. While this could be interpreted in a way which made it conformable with the social scientific endeavour, the reference to human purposes is suggestive of something alien to modern social science, namely, a concern with human values. One of the most extraordinary things about social science and academic work in general is how little it tells us about the question of how we should live. This question obviously involves values and here we immediately run into difficulties. Despite the fact that the practice of science in particular and communication in general depends on commitment to certain values (honesty, sincerity, mutual recognition, etc.), the very forces which have elevated science have also banned it from having anything to do with questions of what is good or bad and what we ought to do (Habermas 1971, 1979, Winch 1979).

Likewise any criticism of society, whether it be of ideas, practices, institutions, or individuals, presupposes the possibility of a better society. Not only this, but criticism implies the *desire* to rid science and human practice in general of ignorance, illusion and constraint on the assumption that freedom from them is a good thing.

Through a long historical process, our conception of values and rationality has been dangerously attenuated; so much so that we find it hard to think of how values might rationally be assessed; to paraphrase Bhaskar, in trying to make science value-free we have made values science-free (Bhaskar 1979). Hence the popular distinction, endorsed by so much of English-speaking philosophy, between matters of fact, which can allegedly be tested decisively, and matters of opinion or belief, which are beyond the scope of rational enquiry. Not only that, values are even seen as a kind of contamination, an archaic fetter upon the development of knowledge. Is it outrageous to suggest that a society which can find no way of deliberating rationally on questions of values, of how we should live, is in deep trouble?

It isn't only that we find it difficult to know how to argue about what is good or bad but that we are reluctant to think about alternative possibilities, good or bad, perhaps because of a legitimate fear of sinking into utopianism and idle speculation. The problem is often exposed by beginning students who, quite reasonably, respond to radical tutors' critical analyses of capitalism or whatever by asking what the solution to the problem is: such questions expose a huge contrast between our eloquence regarding the actual and our embarrassed fumblings regarding our ideas on the possible.

Our difficulties here are reinforced by the 'de-rationalized' nature of values in our society. To echo Cosgrove (this volume), values have customarily come to be thought of as alternately individualistic and universal: in the first case as individual belief or opinion – private, beyond the scope of rational discussion, perhaps even inscrutable; in the second case as lofty, ethereal ideals, divorced from mundane practice but appealed to in the resolution of problems arising from it. Both 'de-socialize' values and divorce them from their function – the harmonization and improvement of human conduct. Moreover, the divorce of values from the social relations and material circumstances to which they pertain is insidious because it enables the ruling class to promote its special interests as the universal interest, as we see in the case of values concerning obedience to authority.

Despite its preoccupation with human values, modern humanistic geography is highly ambiguous about them. We have already noted the reason for this – the misguided assumption that interpretive or humanistic research can abstain from criticizing ideas and meanings in society. Another is its widespread entrapment within that oscillation between individualized and universal conceptions of values identified by Cosgrove. Many 'humanistic geographers' have ironically reinforced these problems. For example, Ted Relph recommends the use of geographical imagination in arguing for change. But while this suggests the possibility of rational discussion between people, the suggestion is promptly subverted by adding the qualifier that one should argue 'according to one's *convictions*', which connotes beliefs which are beyond the scope of rational intersubjective evaluation. This has plenty to do with the widely noted (e.g. Fay 1975) conservatism of some strands of the interpretivist or humanistic movement, with its refusal to evaluate or criticize the beliefs which it seeks to understand, on the false supposition that disagreement and criticism necessarily imply misunderstanding.

Nevertheless, a *critical* humanism can address problems of values, though that this is what it is doing is often overlooked. For example in Raymond Williams' *The country and the city* and *Keywords* the process and results of the contestation of meanings in society are not only interpreted, but interpreted critically, for scarcely a single author in the former book escapes without an objection from Williams. What he shows is that changes in meanings – particularly of words like 'community' – are intimately associated with struggles between different interests. In turn these struggles are not wholly reducible to questions of control over the key material means of life for, as some views of socialism tend to forget, our needs extend beyond these things; they also concern particular kinds of social relations, identity, modes of belonging. As a consequence of this, and as is clear in *The country and the city*, they frequently take the form of a search for a sense of *place*. Furthermore, the contestation of meaning documented in these books is not simply of an academic kind, concerning the best way of conceptualizing something which exists already, outside us. It is also implicitly future-oriented and about values and identity, about notions of an alternative, better way of life – not simply about discovering something that already exists, but about attempts to lead a different life. If many readers of books such as *The country and the city* treat them merely as accounts of the world as an external object of contemplation, then that itself bears witness to the alienation of knowing from being. And if the concern with values seems less than obvious in such literature it is perhaps because the values are neither individualized nor universal and ethereal, but concrete, embedded in historically specific conditions.

Marxism, despite its revolutionary and emancipatory intent, does not escape problems regarding values. Foot, Rigby & Webber (this volume) claim that 'Marxists also apply a political evaluation of theory: can it be used in the long or short term to enhance the power and position of the underprivileged?', but it is rare to see this criterion for evaluating theory being used explicitly in academic Marxist analysis. Marx himself is well known for his contempt for moral questions, yet this hardly fits with his willingness to condemn the iniquities and irrationalities of capitalism and his support for socialism; indeed it is utterly contradictory for Marxists to pretend to have dispensed with questions of values and morality. As Joe McCarney put it: '. . .the price of. . .anti-humanism, with the accompanying loss of every comprehensive image of a better state of human existence, is that we are unable to generate any reasons for caring about the kind of society that Marx spent his life fighting for' (McCarney 1981).

Nor will the criterion of practice do on its own in evaluating theory: many theories can be put into practice successfully without having any emancipatory effect.[5] Further normative criteria are needed which contain some notion of what would constitute a better state of affairs. Of course the danger in the latter argument is that it might open the doors to utopianism and blueprints. This fear is legitimate where proposals are not grounded in existing practice, but all too often it supports a reluctance to say anything of practical use for building a socialist society. The masses will decide, we are told, theory will come from future practice and struggle; but when it comes to it the masses are expected to give a blank cheque to their leaders who themselves have hitherto steadfastly refused to consider what they might do with it. An exaggeration, I know, but it has much to do with the tendency of socialist revolutions to sink into authoritarianism.

Utopianism involves a leap into the future, ignoring the resources and constraints of the present. To avoid utopianism we must locate the seeds of an alternative society within the womb of the present one. But what we have now does not simply consist in inert objects but in potentialities or powers. There is nothing mystical about this; for example, implicit in the concept of labour power is the idea that the potential for labour exists even in someone who is not working now. Such claims are future-oriented but grounded in what exists now. Normally the progressive elements of capitalist society are thought to be its remarkable development of the forces of production, though these only provide necessary, rather than sufficient, conditions for improving the quality of life. However, from the point of view of humanistic Marxism and critical theory, there are also progressive elements in the shape of small areas of democratic and caring practice, which provide a 'draft' or model and throw backward areas into greater relief (Wellmer 1972). Localized advances can be shown from comparison, argument and struggle to require generalization to wider areas of society. Later, through this process, new problems will become apparent, including some in what we used to think of as progressive areas, setting up a different field of tensions.

The problem with this view, particularly at this time, is that reactionary draft meanings favouring the special interests of ruling groups rather than the general interest can also be extended in this way! What distinguishes a humanistic from a reactionary *telos*, however, is that the former reduces exploitation and oppression and can potentially be universalized without simply shifting oppression elsewhere. Or as a critical theorist might put it, reason resists restriction and always demands to be universalized, though it is frequently ignored or overridden in struggle. So in providing a critique of our society Marxism presupposes not only the possibility of a better way, a better life, in which illusions and constraints are pushed back, but the *desire* to achieve these things. To deny this is to render the struggle against oppression and ignorance utterly without point.

Yet despite its radicalism and its greater awareness of the methodological problems of individualizing or universalizing values the left does not escape these tendencies in practice. Sometimes the new right recognizes popular valuations of concrete practices more clearly than the left, as in the case of its appeal to popular discontent with the state. Here it was the left which appealed to lofty values of collectivism, for a long time failing to notice that 'on the ground' people's everyday relationships with the state were frequently demeaning and frustrating.

However, exploring the links between the actual and the potential or possible, is more evident in feminist research. Suzanne Mackenzie's chapter exemplifies this. Unlike orthodox Marxist work, one is given some idea of what women's (and men's) needs are in the present and what they include as regards the immediate future. It is not just that specific needs (e.g. regarding childcare) are mentioned but that there is a stronger sense of agency and becoming, which makes the transition from discussing women's lives to suggesting what they need seem natural rather than requiring a sudden and unaccustomed change of gear. Action is not presented in terms of the reproduction of fixed social relations but in terms of striving to satisfy needs, and hence as future-oriented. This echoes Sartre's critique of Marxist views of agency, which imply that people are driven only by their historical background, not by needs, which are always future-oriented, or by their visions or expectations of their future lives (Sartre 1963; see also Kobayashi, this volume). Accepting these points need not betoken a slide into voluntarism or utopianism: visions of human projects and lives are very much social productions, adapted to material circumstances and conceptualized in socially available terms; and as structurationism reminds us, action presupposes and is carried out within structures, whether acknowledged or not, and even when directed toward different ends, often inadvertently reproduces existing structures. Nevertheless not all action is class or gender struggle. Many men and women's actions are consciously directed toward reproducing the status quo, indeed that this is so is often a necessary condition of the reproduction of social forms such as the nuclear family, rather than an incidental aspect. One of the occupational hazards of radical analysis is underestimating this, so that the reactionary content of people's views of the future is overlooked.

By comparison with feminist analysis, Marxist work on restructuring (including my own!) tends to lack reference to needs and to the relationship between the actual and the possible. (In some cases, developments traditionally thought to make socialism more possible – e.g. the socialization of production – actually seem to make it more remote). Yet perhaps this contrast between Marxist and feminist analysis is not merely due to a different epistemological orientation but has a material basis in the contrast between work in the domestic sphere and work under capitalist social relations. In domestic work there is a much greater proximity of production and consumption; one's feelings about cooking, for example, are closely affected by one's knowledge of how it will be consumed and by whom. By contrast, in the office or the fields or the factory the purpose of work and its relation to needs is obscured. And it is so not simply because of the fact of lack of ownership and control of the means of production but because of the ever-growing complexity and time–space distanciation of the social division of labour. Indeed, I would argue that it is the latter, rather than the former, which presents the principal obstacle to the achievement of democratic control of social production. Marx appears to have thought the same in *The German ideology*.[6] However, that is another argument: the point I want to make here is that it is time we discussed not just the critique of capitalism and patriarchy but alternative, socialist–feminist economics.

Too often the left has been reluctant to work through alternatives which are both feasible and progressive, resorting to the old cop-outs of avoiding utopianism or the dreaded 'reformism'. Alternatively, if the response is that such normative questions are not our task as academics then this needs to be challenged in the light of the above critique of the attenuation of reason. In practice 'New Right' academics have few qualms about devoting time to the evaluation of alternative needs, values and forms of social organization; why should others?

Having made these points about the de-humanization of science and the 'de-rationalization' of ethical deliberations, I have to admit that they do little more than specify the problem. Realist philosophy has made some progress in transcending the divorce of values from science. Bhaskar has shown how value questions are related to explanatory theories and he has theorized the immanent emancipatory character of science; yet in common with so much philosophy of social science there is a curious lack of clarification of what values are (the very term itself may be part of the problem in signifying the attenuation of a wider concept) (Bhaskar 1979, 1986). Part of the problem in much of the literature is the conventional emphasis on the philosophy of science rather than on the philosophy of knowledge and practice in general. If we want to know how knowledge or 'theory' relates to practice, it is fatal to start by reducing the former to science and by treating this in turn as a stock of accumulated knowledge rather than a practice, and then to wonder how all of this relates to everyday practice.[7] Similarly, such starting points must again be resisted if, as Relph and Cosgrove argue, we are to push beyond the divide between science and the arts or humanities.

THEORY, METHOD AND THE PROBLEM OF SYNTHESIS

I now want to turn to some further problems which crop up in many of the foregoing chapters: the role of theory and method in empirical research and the problem of synthesis.

As Ted Relph shows, thinking about the differences and similarities between science and the arts or humanities throws up a contrast, or perhaps a continuum, between 'responsive methods' and 'systematic methods'. The latter are familiar in (so-called) positivist social science, easy to catalogue and formalize into standard procedures, irrespective of subject matter. Yet using methods perfected in physics for the study of people may be as absurd as trying to interview an atom. Disenchantment with systematic procedures has built up steadily over the last two or three decades, not only in reaction to this particular positivist content, but to the whole idea of devising methods independently of the subject matter and purpose of research. It isn't merely a question of adding a few extra types of research method, such as ethnography, to an existing repertoire of systematic methods. It is rather a problem of entering a circle where, to know whether we are asking sensible questions, we have already to know something of the nature of the people and society of whom they are asked, but to know this we must already have asked some questions. Even then, it is not just a matter of adapting methods to be sensitive to those we study; we must also adapt them to ourselves, or rather to the kind of *relationship* we develop with those we study.

Relph argues that 'responsive methods can be clarified, though not reduced to handbook knowledge; they can be rigorously practised'. Two good examples of discussions of responsive methods are the collections edited by Roberts on *Doing feminist research* (Roberts 1981) and by Chambers on research in development studies (Chambers 1982). At the limit, the problem with endorsing responsive methods and refusing to discuss them in abstraction from actual instances of research is that it is difficult to stop questions of method slipping away altogether into a fog of ineffability, which then provides a cover for intellectual complacency and stagnation, masquerading as 'craft'.

It is here that I part company from Relph. Throughout his chapter he equivocates between two conceptions of observation and description: the first is a view of 'pure description' which 'has nothing to do with artifice or technique. . .(or). . .with aesthetics or conception'; the second, a view of observation which involves artistic conventions and which emphasizes the importance of training, practice and technical skills in, for example, 'scales, shading, grammar and punctuation'. An equivalent contrast is that between the now thoroughly discredited view that observation can be theory-neutral and the received view that it is always theory-laden. Relph oscillates between acknowledging the syntactical and semantic structures of perception and description in others – though with the exception of good artists like Turner – and suggesting that we dispense with 'borrowed knowledge' and other cognitive equipment. The myth of unmediated perception in humanistic research is frequently transposed into a characterization of art or literature as a simple expression of beliefs or representation of the world, unmediated by concepts, artistic forms, devices or genres.[8] In this respect, 'the geography of literature and art' is as ignorant of theory in literature and the history of art as humanistic geography generally is of modern philosophy of social science.

However, it is one thing to acknowledge the unavoidability of theory but this then leads to the question of *how* theory is used in research. Several of the chapters centring on Marxist economic analysis focus chiefly on a single theory – that of capital accumulation – in interpreting empirical phenomena. Despite the fact that they address concrete problems, they remain abstract, that is one-sided, in character. Provided we remember this, no harm is done. Such an approach only allows a synthesis in the sense of mining an extensive but thin seam of material, but it invariably presents problems in excavating concrete histories.

Concrete research invariably has to mobilize and integrate concepts from several theories, in order to capture the many-sided nature of its object, even where it is a limited problem, such as the explanation of a migration flow. Synthesis goes one step further, covering a host of such phenomena, though of course even then one's interests are selective. While analysis requires abstraction, synthesis in the sense of 'thick description'[9] requires the development of an ability to command a wide variety of abstractions, theoretical constructs or schemata.

Now many social researchers think of theories as if they were monolithic and discrete and as if incommensurability characterized all inter-theoretical relations. Some philosophers of natural science have reinforced this with their fondness for using formal notation – T', T", etc. – in discussing the relationships between theories, all of which reflects a desperate lack of awareness of the hermeneutical character of these relationships (cf. Bernstein 1976, 1983, Giddens 1976). To undermine these absurd caricatures, we have only to look at the way authors use some parts of major theories and not others and at the way theories are modified unevenly through mediation by other theories of different pedigree. The relationships between humanist Marxists and structuralist Marxists illustrate the situation admirably. Theories overlap, are highly permeable and modifiable, and are only incommensurable with others over limited areas. Consequently, concrete studies or syntheses are not necessarily eclectic in a bad sense.

Seeing theory in this way helps us make sense of 'responsive methods' without imagining that we can abandon all presuppositions and confront situations as a *tabula rasa*. Just as a musician's ability is enhanced by a facility in a large range of different rhythms, scales, chords, etc., and a deep understanding of their structures and interrelationships, so the 'art' of synthesis requires a command over a range of overlapping theories and methods. (The implication in Relph's chapter that we should suspend such knowledge is as absurd as saying that musicians would make better music if they forgot everything they had learned). Yet the theoretical content of a work of synthesis is usually less conspicuous than in works of abstraction because it is plural, complex and nuanced; nevertheless it is still present. The great difficulty is to remain *aware* and critical of the kinds of theoretical mediation that are being used in these circumstances, so that we do not lapse into an unexamined use of everyday concepts – easier said than done!

One possible objection to this line of argument might be that Marxism is not simply a theory but a totalizing science, or, echoing Lukacs, that it is a method. Although Marxism is obviously far more than a method and is unparalleled in its breadth of content and explanatory powers, the first kind of claim invariably merely covers a lack of interest in finding out what else exists in social science.

If synthesis requires the unification of different theories it implies the joining together of the different concerns of humanists and Marxists. As is evident in the foregoing chapters, humanists still tend to be concerned with 'cultural' phenomena, historical materialists with economic phenomena. Such divisions of academic labour are not always harmful but an integration of the cultural and the economic is possible and indeed necessary if we are to achieve *concrete* understandings of society. Here structurationism can offer some guidance in clarifying the way in which one structure can be reproduced by actions directed toward goals relevant to quite different structures. Let us consider some examples concerning an apparently purely economic phenomenon, work under capitalist relations.

Most people who work in this situation tend to work hard, even in the absence of strong managerial pressure or serious threats to their incomes. Sociologists have long been interested in why this should be so. The answer lies in the fact that even when we are hard up we are not motivated solely by economic gain, capitalist ideology or managerial pressure. We are also motivated by the need to maintain respect and co-operation from fellow workers. If one worker slacks, it usually means problems are offloaded on to others, who then withdraw co-operation, respect and civility. Very few of us are immune to such pressures. Though these motivations have origins independent of capital accumulation they help sustain it.

Similarly, Paul Willis' study of why working-class boys get lousy jobs showed that they actively sought out arduous, menial and often poorly paid manual work. Yet they did so not simply through economic coercion but rather through their quest for masculinity and their rejection of authority in school (Willis 1979). Again, Cynthia Cockburn's study of male print workers showed that while trade unions are defined as workers' organizations the craft unions functioned as a medium through which their members constructed and enforced their power as *men* and that this helped the unions' defensive role as labour organizations (Cockburn 1983).

In each of these cases, class relations are reproduced through structures of practices which are partly or wholly directed toward ends not strictly related to class. Conversely, though not reducible to matters of class, these practices presuppose class relations for their execution: without class the workers would not have been in a position to exercise their non-class concerns.

In other words, although we can isolate different structures through abstraction, separating out expressive from economic behaviour for instance, actual concrete processes of structuration involve – and indeed often *require* – the interpenetration and interaction of different structures. Both in the above examples, and in Denis Cosgrove's discussion of the Renaissance, it is clear that a *concrete* understanding of situations such as these has to integrate interpretive/humanistic studies of the expressive order with materialist analyses of the economic order. Such cases demonstrate how such an integration need not involve a lapse into eclectic combinations of inconsistent ideas. More generally they again require us to reject arguments from the humanist side that we should restrict ourselves to actors' accounts and from the structuralist Marxist side that those accounts are of no consequence unless they concern capital and class.

In conclusion, I want to argue that it is *politically* important that we become more accustomed to synthesizing the relationships between the economic and other aspects of society, for in countering reductionism we challenge partial views of political problems.

In part this is a problem of 'indifferent terms', e.g. in describing the state as 'supporting capital accumulation' or 'legitimation'. As an abstraction, state housing *can* be seen as a 'legitimation' of capitalist social relations, but on its own this provides a very limited base for understanding political struggles around housing. The political effects cannot be made intelligible without considering the quality of the use-values and the nature of people's needs. In other words, the political significance of state housing depends not only on the economic aspects of the changes produced in the relationships between state, capital and working class, but also whether, for instance, the housing has adequate heating or whether state housing carries any stigma. Concrete, synthetic studies are needed to appreciate such matters.

Lastly there is a reductionism of a different kind concerning what we mean by capitalist society. In raising this issue I am certainly not questioning that our society is capitalist, but recalling Raymond Williams' point that no economic and political order completely controls or exhausts the content of social life (e.g. sexual behaviour). In overlooking the way in which human endeavour includes quite different purposes, economism tends not to notice the way this can *increase* assent to the prevailing economic order. Moreover economism leads not to the counter-posing of a human order to capitalism but to an alternative way of controlling the division of labour, though of course the latter represents an enormous challenge in itself. Similarly, the reduction of people in capitalism to labour power stunts their development, but some economistic Marxist analyses of capitalism fail to go beyond this definition. As Williams observes: 'It has been the gravest error of socialism, in revolt against class societies, to limit itself so often to the terms of its opponents: to propose a political and economic order, rather than a human order' (Williams 1965).

For this reason alone, the dialogue between humanism and historical materialism is worthy of our attention.

NOTES

1 Cf. Bhaskar's entry on dialectic in Bottomore (1985).
2 Pickles' critique of misconceptions about phenomenology in geography is particularly damning (Pickles 1984).
3 In this context, Relph's claim (this volume) that 'trying to explain. . .is no better than a biologist killing an animal to find out what makes it live' rests upon a wholly fatuous comparison.
4 Cf. Eagleton's farcical attempt to make this reduction in his 1976 *Criticism and ideology.*
5 Not to mention the problem of how we know success or failure is not due to interfering processes.
6 See also Bahro 1978, Meszaros 1987.
7 I tried to avoid this problem in *Method in social science*, particularly in Ch. 1, and hence search for a philosophy of social practice rather than simply a philosophy of social science or scientific practice. While I know I only made very limited progress on this rather grandiose project I must confess to being disappointed that only one reader (a phenomenologist – John Pickles) actually noticed what I was trying to do.
8 Cf. Sayer 1984.
9 Cf. Geertz 1973.

12

Fragmentation, coherence, and limits to theory in human geography

DAVID LEY

The differentiation of science, morality and art has come to mean the autonomy of the segments treated by the specialist and their separation from the hermeneutics of everyday communication (Habermas 1983).

In a persuasive paper, the sociologist Alan Dawe (1973) has contrasted concepts which succeed and those which fail to resonate with everyday human experience. This, I suspect, is a familiar distinction for humanistic geographers. Around 1970 I remember discovering a short manuscript by Yi-Fu Tuan on ' "Environment" and "World" ' (Tuan 1965). In its allusion to human relations, feelings, 'inwardness', even spirituality, this fragment unmasked the representations (rather, the repressions) of human agency in the strongly analytic academic currents of the time. For, in an era of social unrest and experimentation, analytic spatial models did not speak the language of the protests against the Vietnam War, the passion of the civil rights or environmental movements. They failed to engage a reality where the National Guard was marching across the university campus. The lines, of course, will rarely be as sharply drawn, but the lesson is more enduring. An aspiration of humanistic perspectives is to speak the language of human experience, to animate the city and its people, to present popular values as they intersect with the making, remaking and appropriation of place.

In my own attempts to make sense of the experience of daily

life in the black inner city, Tuan's signpost, incomplete as it was, pointed in the direction I was groping for – a human geography that could address the particularities of people and place, and equally intersect broader movements in society. If such an objective appears commonplace today, in the intellectual climate of the early 1970s it was an unlikely venture indeed.

Through the 1970s the contours of such a geography were sketched in, at first experimental and programmatic, later in well-grounded empirical studies. There was considerable diversity as refugees from an analytical (and later a structuralist) perspective which had discarded human agency, made their own, often random, discoveries of Kierkegaard, Merleau-Ponty, Sartre, Mead, Schutz, Raymond Williams and others, to establish an epistemological beach-head. But in this pluralism (which has encouraged considerable misreading by reviewers) there were also common themes, a number of which appeared in the collection *Humanistic geography* (Ley & Samuels 1978). Now around ten years later, it may be worth noting the achievements of humanistic perspectives over a longer period. The first section of this chapter offers a brief retrospective. The second and longer section engages more fully the integrative goals of the present collection. If the first section suggests the contribution of humanistic perspectives to a broader human geography of people and place, it is a moot point to ask whether such a broader vision is attainable. A starting point is a consideration of the much-lamented pluralism in the discipline over the past 20 years. This pattern is interpreted and critiqued against a broader sociology of knowledge which places human geography against the intellectual and cultural history of the 20th century, a period of growing abstraction and specialization of knowledge and speech communities. Against this context, the chapter will end by considering the possibilities for convergence toward a more coherent scholarship.

HUMANISTIC GEOGRAPHY: A BRIEF RETROSPECTIVE

While humanistic studies have a long tradition in historical geography, humanistic emphases within a social science perspective began to appear in the 1960s and early 1970s as a counter to reductionist forms of explanation which denied human and geographical difference. A model of personhood was adopted which upheld the dimension of meaning; human values and experience were integral to a study of people and place. Issues of human agency were central; society consisted of intentional, acting people, and the concepts of subjectivity and intersubjectivity led to the widespread perspective of the social construction of reality, stated concisely by Berger & Luckmann (1966) and later by a number of authors, and most comprehensively by Anthony Giddens (1979). But action, and particularly the culture-building routines of everyday life, are none the less commonly taken for granted and opaque to actors (Ley 1983a). This is one of the formidable obstacles to idealism, that the causes of action (or inaction) are not always apparent to the actors themselves. Expressed motives need to be set in a broader context of opportunities and constraints. An example I have used to make this point is a teenage gang homicide in inner city Philadelphia, where explanation of the event requires an understanding of a range of contexts, from the immanent intersubjective world of inner city youth, to the broader contexts of asymmetric life chances in the American city (Ley 1974, 1980). As such, context is critical. The essays in *Humanistic geography* presented this point forcefully; interpretation passes beyond voluntarism to consider the environment or context in the broadest sense (e.g. Buttimer 1978, Samuels 1978, Western 1978). In short, humanistic geography examines the social construction of place, landscape, or region, as the interplay between people and contexts which they both inherit and help redefine.

This orientation has made some important contributions to the discipline more broadly. In its resistance to a pervasive reductionism and its insistence on a geography with actors, it initiated discussion of human agency, anticipating a central theme in the theoretical discourse of the 1980s. Second, in its depiction of the particularity of place, landscape, and region, it re-established the importance of local difference or geographical contingency. This realization is now widespread, not least in economic geography (Cooke 1986, Gregory & Urry 1985, Massey & Allen 1984) but it encountered great scepticism in the 1970s when featureless space, the isotropic plain, was the postulate of much theoretical geography. A third contribution was the elimination of the fact-value dichotomy, alerting researchers to the irrevocable value-basis of their work. Anne Buttimer's small monograph, *Values in geography* (1974) identified not only the value-impregnated character of local knowledge, but also the value-basis of problem definition, methodology, data retrieval and interpretation. It encouraged greater self-consciousness by scholars, and a self-critical attitude which readily embraced the message of critical theory of an invariable relation between knowledge and social interests. Fourth, the emphasis on values, meaning, and place has proven fertile ground for a revived view of culture and the cultural landscape (Duncan & Ley forthcoming, Kobayashi 1983, Ley 1987, Norton 1987). This has been an area of fruitful meeting among humanists of several stripes, including historical materialists (Cosgrove 1984, Cosgrove & Jackson 1987). A final contribution has been to methodology, with the adoption of a range of qualitative methods, including unstructured interviews and participant observation (Eyles & Donovan 1986, Jackson 1985, Jackson & Smith 1984, Western 1986). These have accompanied a more fully interpretive style of research, geographies of thick rather than thin description (e.g. Daniels 1986, Duncan forthcoming).

But if the contributions of humanistic perspectives to a broader human geography are manifest, so too are its shortcomings. In part they derive from a common flaw in social science, the overspecification of distinctiveness, an exclusive concern with the particularity of a perspective, a form of what Habermas (1983) has called 'the autonomy of the segments'. In the context of humanistic geography this overstatement can lead to a treatment of subjectivity and experience as a virtual fetish, separated from context and material life (Daniels 1985). If an attempt to open up lines for potential integration in human geography is a major objective of this book, then an understanding of disciplinary fragmentation, 'the autonomy of the segments', is a pressing concern. As suggested by the full quotation by Habermas which heads this chapter, fragmentation is a far more general trait of Western society. It is necessary, then, to develop an argument which places the fissiparous tendencies of human geography in the wider context of the cultural history of modernity, including such realms as the arts as well as the social sciences, in order to identify the web of meaning and practice in which the discipline is suspended.

PERVASIVE FRAGMENTATION

First, to establish a baseline, it seems there is more uncertainty about the direction of the discipline today than at any time in the last 20 years. The confidence of, first, the quantitative and, later, the structuralist paradigms has faltered, as a critical and sometimes adversarial scholarship has demonstrated how the excessive claims of an avant-garde have not been attained, and most likely never could be. What are we left with? Few would identify a new consensus, when opinions are so polarized and conflict seems to be so self-evident. Indeed for many practitioners, possibly a majority, confusion would be a more appropriate description of contemporary human geography as a discipline. One recent writer has characterized the field as forming 'a non-relativistic philosophical pluralism' (Paterson 1983, p.175). Another sees the discipline as 'branching toward anarchy' (Johnston 1983, p.189), with 'no consensus, no paradigm-dominance, only a series of mutual accommodations' (p.188). Brian Berry (1980), a former champion of quantitative spatial analysis, is even more apprehensive: 'Pluralism in contemporary geography has gone beyond the creative liberty of variety and degenerated into license that threatens the future of the profession.' The implication seems to be that anything goes, that the doors of a huge theoretical and philosophical emporium have been thrown open, and an uncritical ransacking of the stock is in progress, where the measure of a successful fit is that the product is either suitably new enough or suitably old enough to gain respect. But in this excess human geographers, it seems, are not alone. Daniel Bell (1976, p.13) sees such intellectual libertarianism as a broader trait of modernity: 'Modern culture is defined by this extraordinary freedom to ransack the world storehouse and to engorge any and every style it comes upon.'

It is not difficult to find evidence to support these assessments of the discipline. In 1983 *Geographical Analysis* published a series of invited short essays on the topic of theory in the social sciences. The stimulus to the dozen participants was a brief paper 'Some thoughts about theory in the social sciences' which had appeared in an earlier issue (Papageorgiou 1982), where the author had pondered his subject as he wandered reflectively through a mystical stone garden. The participants (including myself) were given the disarming task of responding to his reverie. Rather like the interpretation of a Rorschach test, the outcome of 'Papa's TAT' was a range of replies among the 12 respondents, which provided a perfect illustration of Gunnar Olsson's claim that theoretical discourse reveals as much or more about the mental categories of the writer as it does about the real world about which he or she is writing. Thus in response to Papageorgiou, Britton Harris (1983) proffered 'I believe that a concern with human health has been one of the driving forces leading to the expansion of biology'; Bob Bennett (1983) saw 'the dilemma between individual and territorial equity' as the central issue, requiring the accounting framework of a negative income tax; Olsson (1983) lisped his way via King Oedipus through 'Expressed impressions of impressed expressions'; Anne Buttimer

(1983) gently and classically invited Wittgenstein to accompany us through 'Teoria, Ryoanji, and the Place Pompidou'; rather grandly, and defensively, I rallied around 'Metaphysics and the place of values in science' (Ley & Pratt 1983); while smarting at real or imagined injuries Peter Gould (1983) took a walk through a vineyard meditating on theory and perversity in the social sciences. The point in all of this was that, given a common stimulus, we took off in a dozen different and non-intersecting directions. What Habermas (1983) saw as the autonomy of the segments has clearly provided further separations within the realm of science itself.

A more serious example of theoretical fragmentation occurs in the edited volume, *A search for common ground* (Gould & Olsson 1982), where, as the title suggests, a self-conscious search for convergence was undertaken. Again there were 12 contributors – one wonders if there is something symbolic about this figure – whose papers were originally presented to a small symposium at a lake retreat in Italy. Those invited represented a cross-section of the discipline, Europeans and North Americans, men and women, graduate students and senior scholars, and representatives of different perspectives in the discipline. Each author was specifically requested to direct his or her thoughts to the search for common ground and in his epilogue Olsson (1982) notes that among the varied contributors 'the closeness of the individual pieces is amazing'. Certainly, the identification of issues shows considerable agreement: the sense, if not for all authors the actuality, of an era emphasizing integration and synthesis, hearkening to the humanist critique for enquiry to pass beyond behaviour to the subjectivity and intersubjectivity of experience, yet conscious of the institutional argument to place experience in a broader explanatory context, a context which may not always be readily visible but is none the less real. There is, then, as Olsson suggests, some ground for optimism for a new disciplinary convergence. But the maturing of this vision into substantial scholarship will depend on the ability of geography's avant-garde to slacken their grasp of the theoretical and methodological vehicles which have brought them along their diverse paths to common ground. Reflecting the state of the discipline, there is a cacophony of theoretical voices in this volume: time geography, structuration, phenomenology, Marxism, Q-analysis, linguistic philosophy, semiotics, each of them one senses holding in part an aesthetic appeal to contributors.

Fragmentation! The same fragmentation as we shall see in other cultural spheres of modernism such as art and architecture. Is the medium the message? Has a form of theoretical elegance, a form of art for art's sake, led some of the authors to their preferred theory and methodology? It is striking how often theory is admired for its beauty, as an aesthetic object. In a presidential address, Leontief (1971) described econometric theory as 'both impressive and in many ways beautiful'. Wittgenstein's biographer detects the same aesthetic in the abstractions of both his philosophy and his house design: '(the house's) beauty is of the same simple and static kind that belongs to the sentences of the *Tractatus*' (Von Wright 1958, p.11). And Charles Jencks (1985) has crisply chastised modern architecture in general for having mistaken pristine beauty for substantive meaning. Is there, perhaps, a symbiotic relationship between the elegance of theory, of a distinctive style, and the self-realization of the intellectual creator? And might the individualism of 'artistic expression' be the most formidable obstacle in the search for common ground?

THEORY AND META-THEORY

Before taking up this question, we should comment briefly on present contenders for paradigm dominance. Several recent writers have identified three: logical positivism, humanism, and a critical position, often but not necessarily a radical political economy (Jackson & Smith 1984, Johnston 1983). While each of the three has its nuances, for present purposes the trichotomy is satisfactory. Logical positivism arrived in geography in the 1960s, and offered valuable standards of rigour and an objective procedure of verifiability through hypothesis testing. But positivism has always been more vulnerable in explanation beyond observable phenomena: humanists identified its weakness in dealing with intangibles, the subjective and intersubjective worlds, and in particular its incapacity to deal with questions of meaning. Structuralists in turn criticized its empiricism, its confinement to an observation language which rendered it unable to account for hidden and underlying causes. Associations might be identified, but not necessarily causal relations, particularly in complex social situations, the very essence of most social science problems. But humanism and structuralism also have their emphases and their blind spots (Daniels 1985, Duncan & Ley 1982).

There have been several attempts to synthesize these three candidates for paradigmatic status into an overarching macro-theory which recognizes the utility, but also the partial insights, of each of them. We might mention three such attempts. The first, Berger & Luckmann's *The social construction of reality* (1966), is an unusually concise and readable volume that, beginning from the starting point of the social individual in everyday life, builds up a more complex portrait of a structured and pre-existing social reality which we inherit, sustain, and in turn modify. A second candidate is the somewhat awesome output of the Frankfurt School, especially Jurgen Habermas (Bernstein 1985, McCarthy 1978), and his identification of the technical, the hermeneutic, and the critical-emancipatory approaches to knowledge, a tripartite set which coincides to some degree with the three contenders for paradigmatic status in human geography. While Habermas' range and scholarship are seminal, it may be fair to state that to the present his argument is both largely abstract and, also, has not achieved a formal integration of the three realms of discourse that he identifies. The third attempt at synthesis is the structuration approach of Anthony Giddens (1979). At present this probably has the greatest currency in theoretical human geography (Gregson 1987, Moos & Dear 1986), and again it contains an awesome breadth of scholarship to which it brings an ambitious attempt at integration. But Giddens' outline is, perhaps, less a thesis than a framework, an inventory of important issues, but issues which are aggregated rather than synthesized. In unsophisticated hands, it can become little more than a checklist, proffering the advice that all things are dialectically related, but without specifying lines of determination.

Each of these attempts at overarching theory has not fully achieved its purposes. But there is a further important point to note here. Each of them represents a second degree abstraction, a theorization of theories, and each is therefore doubly removed from the empirical world. Each encounters substantial problems in integrating, for example, economy and culture, or individual and structure, the so-called problem of human agency, a problem identified by humanists which has become a central preoccupation for theoretical human geography.

But here we must pause and make a striking observation. While *theoretical* integration of such dualisms as agency and structure are demanding, a *practical* resolution of them is made routinely by heavily empirical traditions in the humanities and social sciences and by real human beings every day. The high school history question asking students to contrast immediate and underlying causes of the First World War, for example, directly crosses a chasm which would make theorists tremble. In other historical and cultural disciplines and in the arts, the integration of agency and structure routinely occurs. A novel like *War and peace*, for example, is successful in large measure for its effective interweaving of individuals with their social and historical context. Such endeavours are concerned with telling a story, rather than erecting a theory. These empirical traditions are oriented to issues of understanding and interpretation, and are notable in their disregard of, or even open scepticism toward, questions of theory. One thinks for example of a scholar of the stature of E. P. Thompson and his monumental historiographic work, *The making of the English working class* (1968). But Thompson is also the author of *The poverty of theory* (1978), a skilful assault on certain highly theoretical forms of Marxism. Yet at the same time as he challenges the pursuit of theory, Thompson himself does not reject generalization; it is not the unique *per se* that Thompson or even Tolstoy are interested in.

What is the difference between generalization and theory, and why is it important? Generalization emerges commonly from inductive study and lies closer to the data in its construction of conceptual heuristics; theory in contrast may represent a priori deduction, the imposition of more formal intellectual abstractions upon the data. Francis Bacon characterized this dualism of inductive versus deductive method in a striking metaphor several hundred years ago: 'Empiricists are like ants, they collect and put to use; but rationalists, like spiders, spin threads out of themselves.' Idealism, in this sense, may not be far removed from deductive, theory-building methods. In his critique of structuralism, Thompson observes how the ideas, the a priori categories of the researcher, may be granted a privileged status over empirical circumstances: 'capitalism has become Idea, which unfolds itself in history' (Thompson 1978, p.253). Theory building has become increasingly insulated from the empirical world, in phase with its detachment 'from the hermeneutics of everyday communication'. Increasingly, we see suggestions such as that made regarding central place theory that the theory ultimately may not be testable in that its basic postulates are nowhere met in the real world. As such the theory is admired less for its ability to illuminate reality, than for itself, as an intellectual product with an elegance and coherent logic, with a *beauty* of its own. This is what the autonomy of the segment may imply, a methodological aesthetic, an internalization of scholarship where the intellectual product, most notably theory, is admired for itself. Theory is become not a medium but the message.

How has this come about? And why?

CRITIQUE: RETHINKING THE PLACE OF THEORY

'By our theories you shall know us', the rousing finale of *Explanation in geography* (Harvey 1969, p.486), both reflected and inspired a powerful current of thought in human geography over the past 20 years. The avant-garde in both logical positivism and structuralism, as innovative and vigorous intellectual movements, placed theoretical development at the top of their research agenda. Ian Burton ushered in the quantitative revolution and theoretical geography in 1963 with his remark that 'any branch of geography claiming to be scientific has need for the development of theory' (Burton 1963). This argument was advanced unambiguously by three of the seminal volumes of the 1960s: Bunge's *Theoretical geography* (1966), Chorley & Haggett's *Models in geography* (1967) and Harvey's *Explanation in geography* (1969). By the early 1970s the message had become institutionalized and had diffused rapidly into the more sophisticated introductory texts, where students read statements like 'Theory is the matrix of all science, and the science with an underdeveloped theoretical arm is very much like a ship without a rudder: it drifts rather aimlessly and gets nowhere except by sheer good fortune' (Abler *et al.* 1971, p.45). More recently while positivists have ventured in new directions such as systems theory and information theory, with forays into catastrophe theory and polyhedral dynamics, structuralists have placed theory in at least as privileged a position, and in general have held a sceptical view of empirical research (often denigrated as 'empiricism'), a scepticism which has varied from a criticism of empirical study as preoccupied with superficial appearances, to more severe political charges of its bourgeois and ideological nature. The imperative, in Paul Hirst's words, is 'the necessity of theory' (Hirst 1979). For much innovative work in human geography (and elsewhere), Anne Buttimer is surely right in her suggestion that theory 'became that chameleon-like goddess, who claimed reverence and unquestioning faith' (Buttimer 1983).

There is no need to repeat here the compelling case in favour of a theoretical orientation in social science. The arguments raised in the 1960s are substantial enough: the purely factual nature of idiographic research, its non-cumulative character, the limited search for explanation, all pointed to an incomplete intellectual agenda, which was by no means restricted to human geography. But with hindsight the triumph of theory has been too complete and too non-reflexive, for critical debate on the relative merits of theory and empirical study was essentially stillborn, so that the pursuit of theory has become a given, unreflected and taken for granted. The argument which follows should be clearly understood; while supporting an *orientation* to theory it is critical of a *fixation* upon theory.

Why should this development be seen as problematic? Our word theory comes from the Greek *teoria* or distancing, gaining a separation from the situation at hand. But how much distancing? At what point has a representation of reality been lost? In distancing itself from the idiographic tradition, theoretical endeavour has often achieved a great divorce between its abstractions and the empirical lifeworld, a rupture which we will shortly see finds its affinities in other disciplines and other intellectual fields besides the social sciences. There is an important distinction to be drawn here between the abstractions of theoretical work over the past 20 years, and conceptual heuristics where separation from everyday geographies is more modest.

The problematic distancing of *teoria* from the world we know might be illustrated from David Harvey's work. Following the call to theory in 1969, came a less confident message in 1972: 'There is a clear disparity between the sophisticated theoretical and methodological framework which we are using and our ability to say anything really meaningful about events as they unfold around us' (Harvey 1972b). There began a new engagement with Marxist theory, but ten years later little seems to have changed, as we read in the forthright preface to *The limits to capital*:

> I think it is possible to. . .transcend the seeming boundaries between theory, abstractly formulated, and history, concretely recorded; between the conceptual clarity of theory and the seemingly endless muddles of political practice. But time and space force me to write down the theory as an abstract conception, without reference to history. In this sense the present work is, I fear, but a pale apology for a magnificent conception. And a violation of the ideals of historical materialism to boot. In self-defence I have to say that no one else seems to have found a way to integrate theory and history (Harvey 1982, p.xiv).

What are some of the limitations of a privileged status for theory?

First we find a tendency to instrumentalism, where the empirical record is introduced into the argument only as an adornment or an illustrative device, rather than as a means for generating or formally confirming theoretical claims. The historian Spencer has challenged this use of the historic record, where the theoretician 'loses his openness to the historical-phenomenon-in-itself, and he uses the historical reality to illustrate his theory, thus reversing the desirable relationship of these terms' (Spencer 1977). Raymond Williams is equally critical of a 'theoretistic tendency (which) extends its presumptive interpretations and categories in what is always, essentially, a search for illustrative instances' (Williams 1980, p.34). A cavalier approach to historical conditions is encouraged in a structuralist account with its inbuilt biases against 'empirical appearances' where 'theory is about a theoretical object, and this latter is already several removes from empirical reality' (Glucksmann 1974, p.115). Representation of the empirical world was often treated with disdain by Lévi-Strauss, who upheld the structuralist manifesto that 'the ultimate goal of the human sciences is not to constitute but to dissolve man' (cited in Poster 1978, p.319).

The same tendency is found in econometrics and other highly theoretical areas of economics. In a critique of the loss of empirical representation in economics, Leontief (1971) observed how 'The weak and all too slowly growing empirical foundation cannot support the proliferating superstructure of pure, or should I say, speculative economic theory.' More recently, in an analysis of articles in the *American Economic Review* over almost a five-year period, he noted that a large majority used no data at all, while less than 1% employed data specifically collected by the author to test theoretical propositions (Kuttner 1985). In another guise, some of the humanistic literature is equally incomplete, as eclectic sources are selectively culled to illustrate rather than to test propositions (Daniels 1985).

This raises a related difficulty in the task of verification. If the empirical record is disqualified as an appropriate judge, then how is theory to be verified? Ultimately, as we noted earlier, it was suggested that central place theory may not be testable in as much as its basic postulates were nowhere met in the real world. A similar conundrum faces structuralism and perhaps realism, for in devaluing the empirical world no alternative procedure can convincingly be employed to test the accuracy of theoretical claims. When explanation is referred to deep levels beyond the reach of empirical verification it is possible for self-referential assertions to be presented as self-evident truths. Verification resides in a spiralling internalization of discourse with an ever more elaborate detailing of the purely internal logic of the theory. The result is an ever-increasing abstraction and formalism of discourse until the divorce with the empirical world is complete and reality is made over in a purely formal language. Mystification has occurred, and through the creation of an internalized speech community, theory is sealed against the distractions of what Barbara Kennedy (1979) has called 'a naughty world'.

Access to the community is limited to those who speak the language and belong to an intersubjective social world of shared assumptions. The result can be the elaboration of language games, the pursuit of what Francis Bacon rejected as 'the idols of the tribe'. This anthropological metaphor has been borrowed in a fable by the economist Leijonhufvud (1973) entitled, 'Life among the Econ'. The methodological fetish of the profession can best be understood, he suggests, as tribal rituals: 'Among the Econ . . . status is tied to the manufacture of certain types of implements, called "modls" . . . most of these "modls" seem to be of little or no practical use, [which] probably accounts for the backwardness and abject cultural poverty of the tribe. . .' (cited in Kuttner 1985). As Kuttner demonstrates persuasively, this 'autonomy of the segments' and their 'separation from the hermeneutics of everyday communication' is expressed particularly keenly in economics, a situation of some concern in light of the profession's privileged status in advising public policy.

These trends, then, are by no means limited to human geography, but are present throughout the social sciences and, as we shall see, beyond. In a distinguished lecture to the American Anthropological Association, for example, Flannery (1982) identified a malaise in anthropology associated with the loss of culture as a central paradigm, a divisive fragmentation of the field, and an exalted status for method and theory. There is a striking parallel between the social sciences and the evolution of other intellectual realms in the 20th century including the arts and architecture. Abstraction and the elaboration of intricate codes and protocols have similarly ruptured the link between intellectual constructions in the arts and architecture and representation of the everyday lifeworld. In architectural modernism the built form is essentially the projection of a theory, the imposition of a rationalist abstraction upon historical reality. Local places, emerging slowly and organically, rooted in vernacular traditions and regional cultures, have been systematically erased in favour of a superior architectural logic. Folk knowledge, with its local intricacies, has given way to a universal logic held by the specialist which frequently derides its local constituency. Walter Gropius, an early master of modern architecture, did not consult the workers whose housing he designed, for he regarded them as 'intellectually undeveloped'; his colleague Mies van der Rohe never gave his clients the possibility of choosing among alternative designs:

'How can he choose? He hasn't the capacity to choose' (cited in Knox 1987). Only the specialist possesses the theory, and it is a means of separation from the logics of everyday life.

In a broader sociology of contemporary culture, Bernice Martin has identified in the arts a similar modernist theme 'which focuses on form rather than substantive content', so that 'The pictures (or music) get simpler and barer while the legitimacy theory gets more complex and rococo'. The art exists simply to illustrate the theory, while 'the theory itself becomes the real art work' (Martin 1981, p.88). This underlies Tom Wolfe's complaint that 'these days, without a theory to go with it, I can't see a painting' (Wolfe 1975, p.6), and his invective against modern architecture as the product of academic compounds where theory is exalted over practice (Wolfe 1981). No less than in the social sciences, much of the art of the 20th century has separated itself from the hermeneutics of everyday life, developing its own inaccessible codes and protocols (Rookmaaker 1973). We are reminded of Olsson's remark that abstraction encapsulates the empirical but also the mythical: 'to read urban theory as a text about cities is less penetrating than to approach it as a text of modernity. . .through which both the particular of external reality and the general of internal myth are kept alive' (Olsson 1983). We can see, then, a more general trend emerging where theory as it embraces the mythology of a speech community becomes self-referential and self-confirming.

To the extent that this characterization is accurate it has two further implications that can only be outlined here. First, the acceptance of the notion of the value basis of knowledge means that theoretical positions are both objective and subjective – or, rather, intersubjective, for they indicate a shared meaning system. Like all meaning systems, theories evoke a social world of like-minded others with distinctive social interests. In both senses of the word, theory is ideological. It implicates value systems, even if these are unrecognized and unstated. It is 'pure theory, wanting to derive everything from itself, that succumbs to unacknowledged external conditions and becomes ideological' (Habermas 1971, p.314).

What are the social interests represented by theoretical knowledge? Writers as diverse as Habermas and Bell, as well as a number of geographers, have pointed to the centrality of information and theoretical knowledge in advanced societies. Access to theoretical knowledge is a characteristic of the highly educated, the occupational groups loosely defined as the quaternary sector or the new class. The guild advantage of these groups according to

Gouldner (1979), their cultural capital, resides precisely in their membership of a specialized speech community, securing access to specialized knowledge. Thus theoretical knowledge must be seen not only as idea but also as ideology, supportive of professionalism, of new class interests.

Second, new class ideology (like all ideologies) has political as well as social connotations. To evoke the expert is to evoke a power relationship which may take several forms, such as the conferring of prestige, of influence, or of political authority, each of them implying unequal social relations. Derek Gregory (1980) has noted how systems theory is an ideology of control, and perhaps the best introductory text of a positivist geography concludes with a call to theory in moulding future geographies: 'The various theories of human spatial behavior are as much alternative versions of the world of tomorrow as they are mirrors of the way the world is today. . .our theories are our future' (Abler *et al.* 1971, p.575). This is the deepest dilemma, for if our theories incorporate the reductionism and fragmentation exemplified in theoretical human geography over the past 20 years, then so too will our futures. Theoretical reductionism might well support an existential reductionism, as people are compressed into the matchstick figures that a corporate society models and its social engineers plan for. The suppression of human nature in theory first reflects and then justifies the suppression of humans in practice. We become what we model. And as 'intellectually undeveloped' we are not given the possibility to choose.

POSSIBILITIES FOR INTEGRATION

It is helpful, in short, to see the fragmentation of contemporary human geography as part of a broader expression, not only in social science, but also in the arts, of the cultural codes of modernism. The 'dissolution of man', in Lévi-Strauss' phrase, of human proportion and sensibilities, was accomplished in the arts and architecture as in the social sciences in the search for theoretical universals beyond history and beyond culture, in solutions deeply impregnated by technique, the removal of a regional context and of a recognizable representation of everyday life. The result has been abstraction and the development of private and self-referential codes, inaccessible to a broader public. This is the process Habermas evokes in his commentary on 'the autonomy of the segments treated by the specialist and their separation from the hermeneutics of everyday communication'.

Such abstraction and fragmentation have been criticized before; Leszek Kolakowski (1968) saw them as *The alienation of reason*, and earlier Max Weber, in a related context, also commented critically on 'this fragmentation of the soul' which accompanied the political domination of instrumental reason. In art their argument was anticipated in Goya's etching 'The dream of reason produces monsters', depicting the irrationality of a seemingly rational world.

While remaining cautious to idealistic references to *geist*, there was a cultural movement, a spirit of the age of modernism, with reciprocal lines of contact between the arts, architecture and social science (Ley 1986). There are numerous crossovers: Le Corbusier, the father of modern architecture, was himself a painter and in close contact with avant-garde artists in Europe; he also drew extensively upon the geographer Walter Christaller in his development of urban and regional plans of the built environment in the 1930s (Frampton 1980, p.182). Christaller's view of space coincided with that of the modern architect for whom 'space is seen as isotropic, homogeneous in every direction. . .abstract, limited by boundaries and edges, and rational or logically inferable from part to whole, or whole to part' (Jencks 1981, p.118). Such spaces reveal a broader cultural character of modernism: its theoretical and technical codes were internally logical but self-referential, in part technique for technique's sake, a code shared by a small speech community, but divorced from everyday empirical contexts. There are significant linkages here with the fragmentation and methodological preoccupation we noted earlier as characterizing human geography's *Search for common ground*. At this point, too, we must broaden our definition of modernism to include not only excessive rationalism, the decentring of recognizably human content, but also the oppositional protest of excessive subjectivity expressed by such 20th-century movements as surrealism and some of the post-structuralists. When Gunnar Olsson (1979), who skilfully brings these writers together, concludes one of his papers, 'There are only two of us in the cell; I and my mind', then an absolute threshold has surely been crossed in the self-referential lexicon of the individual self. Might not this be the scientific equivalent of art for art's sake? For abstraction and fragmentation, in art following from the exaltation of technique, of style, in social science contained in the privileged status of theory and method, are related faces of the culture of modernism, whether they represent rational geometries or a highly personal subjectivity.

So we arrive at the fragmentation of self-conscious styles, of intolerant mutual accommodations, of theses and antitheses; but where in this 'autonomy of the segments' can points of synthesis be found? Kenneth Clark, looking at modern art, detected a similar impasse between the abstractions of the competing avant-gardes of rationalism and subjectivity: both 'cubism and super-realism [surrealism], far from being the dawn of a style, are the end of a period of self-consciousness, inbreeding and exhaustion . . .no new style will grow out of a preoccupation with art for its own sake. It can only arise from a new interest in subject matter' (cited in Stern 1980).

If the argument set out here has any merit concerning the interrelatedness of social science with other expressions of modern culture, perhaps contemporary developments in the arts and architecture may contain some clues to a striving toward a new integration in human geography. Post-modernism is a broad-based cultural reaction within the arts and architecture to the abstraction and power of professional elites in modern culture. It is fragile and far from unitary, but the varied strands do have some important points of convergence. First, as intimated in Clark's final sentence there is 'a new interest in subject matter' in arts and architecture, in 'representational as opposed to abstract or conceptual modes' (Jencks 1986, Stern 1980). The attention to specific social and regional contexts, human proportions, and historic traditions in post-modern architecture, for example, all denied by the modern movement, have led in Europe to a revival of traditional vernacular forms (*Architectural Design* 1987, Klotz 1985). In North America, where post-modern architecture is less rooted historically, many design cues are being drawn more eclectically from classical forms. Representation as opposed to abstraction communicates not only to the mind but also to the senses. In their attention to decoration and symbolism, post-modern landscapes aim to be familiar, pleasing, picturesque, even meaningful, speaking unlike modern design to 'the imperatives of the spirit' (Jencks & Chaitkin 1982, p.217). If we might, simplistically, summarize the post-modern project, it is far more fully and self-consciously, to 'make man the measure'. The double coding of post-modern architecture to both the specialist and the user is, in Stern's words, 'an act of communication' to a broader public (in geography, cf. Hasson 1983). Writing of post-modern art more generally, Hutcheon (1986–87) observes how it 'directly engages audiences in the processes of signification. It therefore denies the alienation and transcendence of social milieu that characterized modernism.' In this manner, an attempt is made to restore contact with 'the hermeneutics of everyday communication'.

But this is not to say that the post-modern project has succeeded. Indeed critics abound, though one group of them in particular is primarily defending the earlier separation between the specialist and everyday culture. The challenge to the geographer too is how tenaciously we cling to our 'isms', to a set of abstractions, partial insights and specialized methodological and theoretical vehicles. How easily will we allow the contributions of humanistic geography or historical materialism (or indeed positivism) to become part of an enriched synthesis in geography, and to what extent will we defend our differences, and make a fetish of them, so that like the Econ in Leijonhufvud's fable we proceed toward 'more ornate, ceremonial modls'? At the same time as signs of fragmentation abound, there is also the groping toward a convergence which Gould & Olsson (1982) correctly perceived amid the confusion of tongues, and which this volume also signifies. There are tantalizing possibilities for a renewed human geography, one which is more representational (or empirical) than the recent past, more contextual in its regional, cultural and historical specificity, which seeks to integrate facts and meanings, both the objective and the subjective. There are implications, too, for synthesis as opposed to analysis, for the place of theory which now lies much closer to the ground in conceptual heuristics and lower-order generalizations, and for the style of discourse which reaches beyond the private lexicons of speech communities. Abstraction and fragmentation in human geography are related facets of a broader cultural condition, and as such they should not be confused with a law of nature.

Bibliography

Abler, R., J. Adams & P. Gould 1971. *Spatial organization*. Englewood Cliffs, NJ: Prentice-Hall.

Abrams, P. 1982. *Historical sociology*. Somerset: Open Books.

Ackerman, E. A. 1945. Geographic training, wartime research, and immediate professional objectives. *Annals* Association of American Geographers (A.A.G.) **35**, 121–43.

Ackoff, R. L. 1962. *Scientific method: optimizing applied research decisions*. New York: John Wiley.

Agnew, J. 1981. Home ownership and identity in capitalist societies. In *Housing and identity: cross-cultural perspectives*, J. S. Duncan (ed.), 60–97. London: Croom Helm.

Albrecht, J. & Gill-Chin Lim 1986. Alternative planning theory. *Journal of Architectural and Planning Research* **3**(2), 117–31.

Alexander, E. R. 1984. After rationality, what? A review of responses to paradigm breakdown. *Journal of the American Planning Association* **50**(1), 62–9.

Alinsky, S. 1971. *Rules for radicals*. New York: Random House.

Althusser, L. and E. Balibar 1970. *Reading capital*. London: New Left Books.

Amaral, D. & B. Wisner 1970. Participant observation, phenomenology and the rules for judging sciences: a comment. *Antipode* **2**(1), 42–51.

American Iron and Steel Institute 1983. *Annual Statistical Report*. Washington D.C.: American Iron and Steel Institute.

Anderson, J. 1973. Ideology in geography: an introduction. *Antipode* **5**(3), 4–6.

Anderson, P. 1974. *Lineages of the absolutist state*. London: New Left Books.

Appleton, J. 1975a. Landscape evaluation: the theoretical vacuum. *Transactions of the Institute of British Geographers* **66**, 120–33.

Appleton, J. 1975b. *The experience of landscape*. New York: John Wiley.

Architectural Design, 1987. Post-modernism and discontinuity, **57** (1–2), whole issue.

Armstrong, A. H. 1968. Thomas Adams and the Commission of Conservation. In *Planning the Canadian environment*, L. O. Gertler (ed.), 17–35. Montreal: Harvest House.

Armstrong, P. & H. Armstrong 1983. Beyond numbers: problems with quantitative data. *Alternate Routes* **6**, 1–40.

Armstrong, W. R. and T. G. McGee 1985. *Arenas of accumulation*. London: Methuen.

Arnstein, S. R. 1969. A ladder of citizen participation. *Journal of the American Institute of Planners* **35**, 216–24.

Aron, R. 1975. *History and the dialectic of violence: an analysis of Sartre's 'Critique de la Raison Dialectique'*. Oxford: Blackwell.

Atkins, L. & D. Jarrett 1979. The significance of 'significance tests'. In *Demystifying social statistics*, J. Irvine, I. Miles & J. Evans (eds), 87–109. London: Pluto Press.

Bahro, R. 1978. *The alternative in Eastern Europe*. London: New Left Books.

Barbieri, G. 1983. *Andrea Palladio e la Cultura Veneta del Rinascimento*. Rome: Veltro.

Barnes, T. J. & M. G. Curry 1983. Towards a contextualist approach in human geography. *Transactions of the Institute of British Geographers* NS **8**, 467–82.

Barnett, D. 1977. *Comparative hourly labour costs in the world steel industry*. Ottawa: Resource Industries and Construction Branch, Department of Industry, Trade and Commerce.

Barnett, D. & L. Schorsch 1983. *Steel: upheaval in a basic industry*. Cambridge, Mass.: Ballinger.

Barrett, M. 1980. *Women's oppression today: problems in Marxist feminist analysis*. London: Verso.

Baxandall, M. 1972. *Painting and experience in fifteenth century Italy*. Oxford: Clarendon Press.

Beauregard, R. 1980. Teaching planning theory: dilemmas beyond the rational model. *Bulletin of the Association of Collegiate Schools of Planning* **18**(1), 1–4.

Bell, D. 1976. *The cultural contradictions of capitalism*. New York: Basic Books.

Bennett, R. J. 1983. Individual and territorial equity. *Geographical Analysis* **15**, 50–57.

Benton, T. 1977. *The three sociologies*. London: Routledge & Kegan Paul.

Berger, P. & T. Luckmann 1966. *The social construction of reality*. Garden City, N.Y.: Doubleday.

Berka, K. 1983. *Measurement: its concepts, theories and problems*. (Trans. A. Riska.) Dordrecht and Boston: Reidel.

Bernstein, R. J. 1976. *The restructuring of social and political theory*. Oxford: Blackwell.

Bernstein, R. J. 1983. *Beyond objectivism and relativism*. Oxford: Blackwell.

Bernstein, R. J. (ed.) 1985. *Habermas and modernity*. Cambridge: Polity Press.

Berry, B. 1972. More on relevance and policy analysis. *Area* **4**(2), 77–80.

Berry, B. 1980. Creating future geographers. *Annals A.A.G.* **70**, 449–58.

Betheil, R. 1978. The ENA in perspective: the transformation of collective bargaining in the basic steel industry. *The Review of Radical Political Economics* **10**(2), 1–25.

Bhaskar, R. 1975. *A realist theory of science*. Brighton, Sussex: Harvester Press.

Bhaskar, R. 1979. *The possibility of naturalism*. Brighton, Sussex: Harvester Press.

Bhaskar, R. 1986. *Scientific realism and human emancipation*. London: Verso.

Billinge, M. 1977. In search of negativism: phenomenology and historical geography. *Journal of Historical Geography* 3(1), 55–67.

Billinge, M. 1983. The Mandarin dialect. *Transactions of the Institute of British Geographers* NS **8**, 400–420.

Blake, W. n.d. *Jerusalem*.

Blau, J. R., M. E. LaGory & J. S. Pipkin (eds) 1983. *Professionals and urban form*. Albany: State University of New York Press.

Blowers, A. 1972. Bleeding hearts and open valves. *Area* 4(4), 290–92.

Blumenfeld, H. 1967. *The modern metropolis*. (Ed. Paul D. Spreiregen.) Montreal: Harvest House.

Boland, L. A. 1982. *The foundations of economic method*. London: George Allen & Unwin.

Bosquet, M. 1977. *Capitalism in crisis and everyday life*. Hassocks: Harvester.

Bottomore, T. (ed.) 1985. *A dictionary of Marxist thought*. Oxford: Blackwell.

Bourne, L. S. (ed.) 1971. *Internal structure of the city: readings on space and environment*. New York and London: Oxford University Press.

Bowlby, S., J. Foord, & S. Mackenzie 1981. Feminism and geography. *Area* **13**(4), 711–16.

Bradbury, B. 1982. The fragmented family: family strategies in the face of death, illness and poverty, Montreal, 1860–1885. In *Childhood and family in Canadian history*, J. Parr (ed.), 109–28. Toronto: McClelland & Stewart.

Bradbury, J. H. 1984. The impact of industrial cycles in the mining sector: the case of the Quebec-Labrador region in Canada. *International Journal of Urban and Regional Research* **8**, 311–31.

Bradbury, J. H. 1985. Regional and industrial restructuring in the international division of labour. *Progress in Human Geography* **9**(1), 38–63.

Bradbury, J. H. & I. St Martin 1983. Winding-down in a Quebec mining town: A case study of Schefferville. *The Canadian Geographer* **27**, 128–44.

Branch, M. C. (ed.) 1975. *Urban planning theory*. Stroudsburg, Pennsylvania: Dowden Hutchinson & Ross (Community Development Series, 15).

Branch, M. C. 1983. *Comprehensive planning: general theory and principles*. Pacific Pallisades, Ca.: Pallisades Press.

Braybrooke, D. & C. E. Lindblom 1963. *A strategy of decision*. New York: Free Press.

Breines, W. 1982. *Community and organization in the New Left 1962–1968*. New York: Praeger.

Breitbart, M. 1972. Advocacy in planning and geography. *Antipode* 4(2), 64–8.

Bridenthal, R. 1976. The dialectic of production and reproduction in history. *Radical America* **10**(3), 3–11.

Brown, B. 1973. *Marx, Freud and the critique of everyday life: toward a permanent cultural revolution*. New York: Monthly Review.

Bunge, W. 1966. *Theoretical geography*. Lund: Gleerup.

Bunge, W. 1977. The first years of the Detroit geographical expedition: a personal report. In *Radical geography*, R. Peet (ed.), 31–40. Chicago: Maaroufa.

Burawoy, M. 1979. *Manufacturing consent: changes in the labour process under monopoly capitalism*. Chicago: Chicago University Press.

Burchell, R. W. & G. Sternlieb (eds) 1978. *Planning theory in the 1980s: a search for future directions*. New Brunswick, NJ: Rutgers.

Burgess, R. 1976. *Marxism and geography*. Occasional papers, Department of Geography, University College, London, 30.

Burton, I. 1963. The quantitative revolution and theoretical geography. *Canadian Geographer* **7**, 151–62.

Business Week 1982. US steel workers slim down for survival. 31 May, 88–9.

Buttimer, A. 1974. Values in geography. Commission on College Geography, Resource Paper 24. Washington: Association of American Geographers.

Buttimer, A. 1978. Charisma and context: the challenge of la géographie humaine. In *Humanistic geography: prospects and problems*. D. Ley & M. Samuels (eds), 58–76. Chicago: Maaroufa.

Buttimer, A. 1983. Teoria, Ryoanji, and the Place Pompidou. *Geographical Analysis* **15**, 42–6.

Canada Central Mortgage and Housing Corporation 1969. *Kingston housing survey*. Ottawa: Central Mortgage and Housing Corporation.

Canada National Health and Welfare, National Day Care Information Centre 1984. *Status of day care in Canada, 1983*. Ottawa: National Health and Welfare.

Carchedi, G. 1977. *On the economic identification of social classes*. London: Routledge & Kegan Paul.

Carlyle, T. 1840. *On heroes, hero-worship and the heroic in history*. Extracts reprinted in F. Stern (ed.) 1982. *The varieties of history*. New York: Vintage.

Carr, E. H. 1964. *What is history?* Harmondsworth: Penguin.

Carson, R. 1962. *Silent spring*. Boston: Houghton Mifflin.

Carter, H. 1962. *The study of urban geography*. London: Edward Arnold.

Cassirer, E. 1964. *The individual and the cosmos in Renaissance philosophy*. New York: Harper & Row.

Chadwick, G. 1971. *A systems view of planning*. Oxford: Pergamon.

Chambers, R. 1982. *Rural development: putting the last first*. London: Longman.

Chapin, F. S. 1965. *Urban land use planning*. Urbana, Ill.: University of Illinois Press.

Chesler, P. 1972. *Women and madness*. New York: Avon.

Chiodi, P. 1978. *Sartre and Marxism*. Hassocks, Sussex: Harvester Press.

Chisholm, M. 1971. Geography and the question of relevance. *Area* **3**(2), 65–8.

Chorley, R. & P. Haggett (eds) 1967. *Models in geography*. London: Methuen.

Chouinard, V. & R. Fincher 1983. A critique of 'Structural Marxism and Human Geography'. *Annals*, A.A.G. **73**(1), 137–46.

Chouinard, V., R. Fincher & M. Webber 1984. Empirical research in scientific human geography. *Progress in Human Geography* **8**(3), 347–80.

Clavel, P., J. Forester & W. W. Goldsmith (eds) 1980. *Urban and regional planning in an age of austerity*. Oxford: Pergamon.

Cloward, R. A. & F. F. Piven 1979. *Poor people's movements*. New York: Random House.

Cockburn, C. 1983. *Brothers*. London: Pluto.

Cohen, L. J. 1970. *The implications of induction*. London: Methuen.

Cohen, S. S. 1977. *Modern capitalist planning: the French model*. Berkeley: University of California Press.

Collins, J. 1968. *The Existentialists: a critical study*. Chicago: Henry Regnery Company.

Commoner, B. 1971. *The closing circle: nature, man and technology*. New York: Knopf.

Cooke, P. 1986. The changing urban and regional system in the UK. *Regional Studies* **20**, 243–51.

Coote, A. & P. Kellner 1980. *Hear this brother: women workers and union power*. London: New Statesman.

Corey, K. 1972. Advocacy in planning: a reflective analysis. *Antipode* **4**(2), 46–63.

Cornish, V. 1935. *Scenery and the sense of sight*. Cambridge, Mass.: Harvard University Press.

Cosgrove, D. 1978. Place, landscape and the dialectics of cultural geography. *The Canadian Geographer* **22**(1), 66–71.

Cosgrove, D. 1983. Towards a radical cultural geography: problems of theory. *Antipode* **15**(1), 1–11.

Cosgrove, D. 1984. *Social formation and symbolic landscapes*. London: Croom Helm.

Cosgrove, D. 1985. Prospect, perspective and the evolution of the landscape idea. *Transactions of the Institute of British Geographers* NS **10**(1), 43–62.

Cosgrove, D. & P. Jackson 1987. New directions in cultural geography. *Area* **19**, 95–101.

Couclelis, H. & R. Golledge 1983. Analytic research, positivism and behavioural geography. *Annals*, A.A.G. **73**(3), 331–9.

Crandall, R. 1981. *The US steel industry in recurrent crisis*. Washington: Brookings Institute.

Crooks, H. 1983. *Dirty business*. Toronto: Lorimer.

Curry, M. G. 1980. *Forms of life: a Wittgensteinian view*. Unpublished MA thesis, University of Minnesota.

Curry, M. G. 1985. *Universal and local models in urban geography*. Paper presented at the 81st Annual Meeting of the Association of American Geographers, Detroit, Michigan, April.

Cutler, A. I., B. Hindess, P. Q. Hirst & A. Hussain 1977. *Marx's capital and capitalism today 1.* London: Routledge & Kegan Paul.

Daniels, S. 1985. Arguments for a humanistic geography. In *The future of geography*, R. Johnston (ed.), 143–58. London: Methuen.

Daniels, S. 1986. The implications of industry: Turner and Leeds. *Turner Studies* **6**(1), 10–17.

Davidoff, P. 1965. Advocacy and pluralism in planning. *Journal of the American Institute of Planners* **31**, 331–8.

Dawe, A. 1973. The role of experience in the construction of social theory. *Sociological Review* **21**, 25–55.

Dear, M. & A. Scott 1981. Towards a framework for analysis. In *Urbanization and urban planning in capitalist society*, M. Dear & A. Scott (eds), 3–16. London and New York: Methuen.

de Beauvoir, S. 1961. *The second sex.* (Trans. and ed. O. Parshley.) New York: Bantam.

Debus, A. 1978. *Man and nature in the Renaissance.* Cambridge: Cambridge University Press.

Dee, J. 1570. *The mathematicall praeface to the elements of geometrie of Euclide of Megara.* (Reprinted with introduction by Allen G. Debus 1975.) New York: Science History Publications.

Dobb, M. H. 1940. *Political economy and capitalism.* New York: International Publishers.

Dobb, M. H. 1973. *Theories of value and distribution since Adam Smith.* Cambridge: Cambridge University Press.

Doeringer, P. B. & M. J. Piore 1971. *Internal labour markets and manpower analysis.* Lexington, Mass.: Heath Lexington Books.

Dumont, L. 1977. *From Mandeville to Marx: the genesis and triumph of economic ideology.* Chicago: Chicago University Press.

Duncan, J. 1980. The superorganic in American cultural geography. *Annals, A.A.G.* **70**(2), 181–98.

Duncan, J. 1985. Individual action and political power: a structuration perspective. In *The future of geography*, R. J. Johnston (ed.), 174–89. New York: Methuen.

Duncan, J. forthcoming. *The city as text: the landscape of charismatic rule in pre-colonial Kandy.* Cambridge: Cambridge University Press.

Duncan, J. & N. Duncan 1984. A cultural analysis of urban residential landscapes in North America: the case of the anglophile elite. In *The city in cultural context*, J. Agnew, J. Mercer & D. Sopher (eds), 255–76. Boston: Allen & Unwin.

Duncan, J. S. & D. F. Ley 1982. Structural Marxism and human geography: a critical assessment. *Annals A.A.G.* **72**(1), 30–59.

Duncan, J. S. & D. F. Ley 1983. Comment in reply. *Annals A.A.G.* **73**(1), 146–50.

Duncan, J. S. & D. F. Ley forthcoming. Cultural geography. In

Geography in the social sciences, G. Smith, D. Gregory, & R. Martin (eds). London: Macmillan.

Dunleavy, P. 1979. The urban basis of political alignment: social class, domestic property, and state intervention in consumption processes. *British Journal of Political Science* **9**, 409–43.

Eagleton. T. 1976. *Criticism and ideology*. London: New Left Books.

Edel, M. 1981. Capitalism, accumulation and the explanation of urban phenomena. In *Urbanization and urban planning in capitalist society*, M. Dear & A. Scott (eds), 19–44. London and New York: Methuen.

Edwards, R. 1979. *Contested terrain: the transformation of the workplace in the twentieth century*. London: Heinemann.

Edwards, R., M. Reich & M. Gordon (eds) 1975. *Labour market segmentation*. Lexington, Mass.: D.C. Heath.

Eisenstein, Z. (ed.) 1979. *Capitalist patriarchy and the case for socialist feminism*. New York: Monthly Review Press.

Ellul, J. 1954. *The technological society*. New York: Knopf.

Engels, F. 1934. *Dialectics of nature*. Moscow: Progress.

Entrikin, J. N. 1976. Contemporary humanism in geography. *Annals A.A.G.* **66**(4), 615–32.

Ernst, R., R. Crooker & R. Ayott 1972. Competition and conflict over land use: change in the inner city: institution versus community. *Antipode* **6**(2), 70–97.

Etzioni, A. 1967. Mixed scanning: a 'third' approach to decision making. *Public Administration Review* **27**, 385–92.

Evernden, N. 1985. *The natural alien: humankind and environment*. Toronto: University of Toronto Press.

Eyles, J. 1981. Why geography cannot be Marxist: toward an understanding of lived experience. *Environment and Planning A* **13**, 1371–88.

Eyles, J. & J. Donovan 1986. Making sense of sickness and care: an ethnography of health in a West Midlands town. *Transactions of the Institute of British Geographers NS* **11**, 415–27.

Faludi, A. (ed.) 1973a. *A reader in planning theory*. New York: Pergamon.

Faludi, A. 1973b. *Planning theory*. New York: Pergamon.

Fann, F.T. 1971. *Wittgenstein's conception of philosophy*. Berkeley: University of California Press.

Fava, S. 1980. Women's place in the new suburbia. In *New space for women*, G. Wekerle, R. Peterson & D. Morley (eds), 129–50. Boulder: Westview Press.

Fay, B. 1975. *Social theory and political practice*. London: George Allen & Unwin.

Firestone, S. 1971. *The dialectic of sex*. New York: Bantam.

Fitzgerald, M., C. Guberman & M. Wolfe (eds) 1982. *Still ain't satisfied: Canadian feminism today*. Toronto: The Women's Press.

Flannery, K. 1982. The golden Marshalltown: a parable for the archeology of the 1980s. *American Anthropologist* **84**, 265–78.

252   *Bibliography*

Foreman, A. 1977. *Femininity as alienation: women and the family in Marxism and psychoanalysis.* London: Pluto.

Forester, J. 1980. Critical theory and planning practice. *Journal of the American Planning Association* **46**, 275–86.

Forester, J. 1982. Know your organisations: planning and the reproduction of social and political relations. *Plan Canada* **22**(1), 3–13.

Fosberg, S. 1981. *Family day care in the United States: summary of findings.* Final Report of the National Day Care Home Study, Volume 1. Washington, DC: United States Department of Health and Human Services.

Foucault, M. 1969. *L'Archéologie du Savoir.* Paris: Gallimard. (Trans. 1972.) *The archeology of knowledge.* New York: Pantheon.

Fox, B. 1980. *Hidden in the household: women's domestic labour under capitalism.* Toronto: The Women's Press.

Frampton, K. 1980. *Modern architecture: a critical history.* New York: Oxford University Press.

French, P. 1972. *John Dee: The world of an Elizabethan magus.* London: Routledge & Kegan Paul.

Friedmann, J. 1973. *Retracking America: a theory of transactive planning.* Garden City, NY: Anchor Books.

Friedmann, J. & B. Hudson 1974. Knowledge and action: a guide to planning theory. *Journal of the American Institute of Planners* **40**, 2–16.

Friendly, M. & L. Johnson 1981. *Perspectives on work-related day care,* Child in the City 11. Centre for Urban and Community Studies, University of Toronto.

Froebel, F., J. Heinrichs & O. Kreye 1980. *The new international division of labour.* London: Cambridge University Press.

Fromm, E. 1962. *Beyond the chairs of illusion: my encounter with Marx and Freud.* New York: Simon & Schuster.

Gallion, A. B. & S. Eisner 1950. *The urban pattern.* New York: Van Nostrand.

Garin, E. 1969. *Science and civil life in the Italian Renaissance.* Gloucester, Massachusetts.

Geddes, P. 1915. *Cities in evolution.* London: Williams & Norgate.

Geertz, C. 1973. *The interpretation of cultures.* New York: Basic Books.

Gellner, E. 1973. *Cause and meaning in the social sciences.* London: Routledge & Kegan Paul.

Gershuny, J. 1978. *After industrial society: the emerging self-service economy.* London: Macmillan.

Gibson, K. D. & R. J. Horvath 1983a. Aspects of a theory of transition within the capitalist mode of production. *Environment and Planning D: Society and Space* **1**, 121–38.

Gibson, K. D. & R. J. Horvath 1983b. Global capital and the restructuring crisis in Australian manufacturing. *Economic Geography* **51**, 178–94.

Giddens, A. 1976. *New rules of sociological method: a positive critique of interpretative sociologies.* London: Hutchinson.

Giddens, A. 1977. *Studies in social and political theory.* London: Hutchinson.

Giddens, A. 1979. *Central problems in social theory.* Berkeley: University of California Press.

Giddens, A. 1981. *A contemporary critique of historical materialism.* Volume 1, *Power, property and the state.* London: Macmillan.

Giddens, A. 1984. *The constitution of society.* Cambridge: Polity Press.

Giere, R. 1980. Statistical hypothesis. In *Applications of inductive logic,* L. J. Cohen & M. Hesse (eds), 251–70. Oxford: Clarendon Press.

Giustiniani, V. R. 1985. Home, humanus, and the meanings of 'humanism'. *Journal of the History of Ideas* **46**(2), 167–95.

Glucksmann, M. 1974. *Structuralist analysis in contemporary social thought.* London: Routledge & Kegan Paul.

Gold, J. & J. Burgess (eds) 1982. *Valued environments.* London: Allen & Unwin.

Goldmann, L. 1969. *The human sciences and philosophy.* London: Jonathan Cape.

Goodman, R. 1971. *After the planners.* New York: Simon & Schuster.

Gordon, D. M., R. Edwards & M. Reich 1982. *Segmented work, divided workers. The historical transformation of labour in the United States.* Cambridge: Cambridge University Press.

Gould, P. 1970. Is statistix inferens the geographical name for a wild goose? *Economic Geography* **4**, 439–48.

Gould, P. 1983. On the road to Colonus: or theory and perversity in the social sciences. *Geographical Analysis* **15**, 35–40.

Gould, P. & G. Olsson (eds) 1982. *A search for common ground.* London: Pion.

Gouldner, A. W. 1979. *The future of intellectuals and the rise of the new class.* New York: Seabury.

Gouldner, A. W. 1980. *The two Marxisms.* New York: Seabury.

Government Accounting Office 1974. *Economic and foreign policy effects of voluntary restraint agreements on textiles and steel.* Washington: US Government Printing Office.

Government Accounting Office 1980. *Administration of the steel trigger price mechanism.* Washington: US Government Printing Office.

Grayson, J. P. 1985. *Corporate strategy and plant closures. The SKF experience.* Toronto: Our Times Press.

Greenberg, D. & R. Walker 1982. Post-industrialism and political reform in the city: a critique. *Antipode,* **14**(1), 17–32.

Greer, G. 1971. *The female eunuch.* London: Paladin.

Gregory, D. 1978. *Ideology, science and human geography.* London: Hutchinson.

Gregory, D. 1980. The ideology of control: systems theory and geography. *Tijdschrift voor Economische en Social Geografie* **71**, 327–42.

Gregory, D. 1981. Human agency and human geography. *Transactions of the Institute of British Geographers* NS **6**, 1–18.

Gregory, D. 1982. Solid geometry: notes on the recovery of spatial structure. In *A search for common ground*, P. Gould & G. Olsson (eds), 187–219. London: Pion.

Gregory, D. 1984. Space, time and politics in social theory: an interview with Anthony Giddens. *Environment and Planning D: Society and Space* **2**, 119–32.

Gregory. D. & J. Urry (eds) 1985. *Social relations and spatial structures*. London: Macmillan.

Gregson, N. 1986. On duality and dualism: the case of structuration and time geography. *Progress in Human Geography* **10**(2), 184–205.

Gregson, N. 1987. Structuration theory: some thoughts on the possibilities for empirical research. *Society and Space* **5**, 73–91.

Griffiths, D., J. Irvine & I. Miles 1979. Social statistics: towards a radical science. In *Demystifying social statistics*, J. Irvine, I. Miles & J. Evans (eds) 339–81. London: Pluto Press.

Gudeman, S. 1978. *The demise of a rural economy*. London: Routledge & Kegan Paul.

Gudeman, S. 1982. *Ricardo's representations*. Unpublished paper, University of Minnesota.

Guelke, L. 1974. An idealist alternative in human geography. *Annals A.A.G.* **64**, 193–202.

Gunton, T. I. 1981. *The evolution and practice of urban and regional planning in Canada 1900–1960*. Unpublished PhD dissertation. University of British Columbia.

Habermas, J. 1971. *Knowledge and human interests*. London: Heinemann.

Habermas, J. 1979. *Communication and the evolution of society*. Boston: Beacon Press.

Habermas, J. 1983. Modernity: an incomplete project. In *The anti-aesthetic: essays on postmodern culture*, H. Foster (ed.), 3–15. Port Townsend, Wash.: Bay Press.

Hackett, R. A. 1980. Pie in the sky. A history of the Ontario Waffle. *Canadian Dimension* **15** (1, 2), 1–72.

Hacking, I. 1965. *Logic of statistical inference*. Cambridge: Cambridge University Press.

Hahn, F. 1983. The neo-Ricardians. *Cambridge Journal of Economics* **6**, 353–74.

Hall, J. M. 1982. *The geography of planning decisions*. Oxford: Oxford University Press.

Hamilton, R. & M. Barrett (eds) 1986. *The politics of diversity: feminism, Marxism and Canadian society*. London: Verso.

Harcourt, G. C. 1972. *Some Cambridge controversies in the theory of capital*. Cambridge: Cambridge University Press.

Harcourt, G. C. 1982. The Sraffan contribution: an evaluation. In *Classical and Marxian political economy*, I. Bradley & M. Howard (eds), 255–75. London: Macmillan.

Harloe, M. (ed.) 1977. *Captive cities: studies in the political economy of cities and regions*. London: John Wiley.

Harré, R. 1970. *The principles of scientific thinking*. London: Macmillan.

Harré, R. 1972. *The philosophies of science.* Oxford: Oxford University Press.

Harré, R. 1979. *Social being.* Oxford: Blackwell.

Harris, B. 1983. Interdisciplinary disciplines. *Geographical Analysis* **15**, 47–50.

Harris, C. 1971. Theory and synthesis in historical geography. *The Canadian Geographer* **15**(3), 157–72.

Harris, C. 1978. The historical mind and the practice of geography. In *Humanistic geography: prospects and problems*, D. Ley & M. Samuels (eds), 123–37. Chicago and London: Maaroufa/Croom Helm.

Harris, R. 1984. A political chameleon: class segregation in Kingston, Ontario, 1961–1976. *Annals A.A.G.* **74**, 454–76.

Harris, R. 1985. The new left in urban politics. *Queen's Quarterly* **92**(3), 572–89.

Harris, R. 1988. *Democracy in Kingston: a social movement in urban politics, 1965–1970.* Kingston and Montreal: McGill-Queen's University Press.

Hartshorne, R. 1939. *The nature of geography.* Lancaster: Association of American Geographers.

Harvey, D. 1969. *Explanation in geography.* London: Edward Arnold.

Harvey D. 1972a. Revolutionary and counter-revolutionary theory in geography and the problem of ghetto formation. *Antipode* **4**(2), 1–13, 36–41.

Harvey, D. 1972b. Revolutionary and counter-revolutionary theory in geography and the problem of ghetto formation. In *Geography of the ghetto*, H. Rose (ed.), 1–25. DeKalb, Ill.: Northern Illinois University Press.

Harvey, D. 1973. *Social justice and the city.* London: Edward Arnold.

Harvey, D. 1974. What kind of geography for what kind of public policy? *Transactions of the Institute of British Geographers* **63**, 18–24.

Harvey, D. 1975. The geography of capital accumulation: a reconstruction of the Marxist theory. *Antipode* **7**(2), 9–21.

Harvey, D. 1978. Labour, capital and class struggle around the built environment in advanced capitalist societies. In *Urbanization and conflict in market societies*, K. Cox (ed.), 9–37. Chicago: Maaroufa.

Harvey, D. 1979. Monument and myth. *Annals A.A.G.* **69**(3), 362–81.

Harvey, D. 1982. *The limits to capital.* Oxford: Blackwell.

Harvey, D. 1984. On the history and present condition of geography: an historical materialist manifesto. *The Professional Geographer* **35**, 1–10.

Harvey, D. 1987. Three myths in search of a reality. *Society and Space: Environment and Planning D.*

Hasson, S. 1983. *The neighbourhood organization as a pedagogic project.* Institute of Urban and Regional Studies, Hebrew University of Jerusalem.

Heidegger, M. 1966. *Discourse on thinking.* New York: Harper & Row.

Higgins, D. J. H. 1977. *Urban Canada. Its government and politics.* Toronto: Macmillan.

Hightower, H. C. 1969. Planning theory in contemporary professional education. *Journal of the American Institute of Planners* **35**, 326–9.

Hindess, B. 1977. *Philosophy and methodology in the social sciences.* Hassocks: Harvester Press.

Hiorns, F. R. 1956. *Town building in history.* London: Harrap.

Hirsh, A. 1981. *The French new left: an intellectual history from Sartre to Gorz.* Boston: South End Press.

Hirst, P. 1979. The necessity of theory. *Economy and Society* **8**, 417–45.

Hoch, C. 1984. Doing good and being right: the pragmatic connection in planning theory. *Journal of the American Planning Association* **50**(3), 335–45.

Hodge, G. 1986. *Planning Canadian communities: an introduction to the principles, practices and participants.* Toronto: Methuen.

Holloway, J. & S. Picciotto (eds) 1978. *State and capital: a Marxist debate.* London: Edward Arnold.

Holt-Jensen, A. 1980. *Geography: its history and concepts.* London: Harper & Row.

Horvath, R. 1970. On the relevance of participant observation. *Antipode* **2**(1), 30–37.

Howard, E. 1898. *Tomorrow: a peaceful path to real reform.* (Reissued as *Garden cities of tomorrow,* 1902 reprinted.) London: Faber & Faber. (1946 edn.)

Howe, E. & J. Kaufman 1979. The ethics of contemporary American planners. *Journal of the American Planning Association* **43**(3), 243–55.

Hughes, J. A. 1980. *The philosophy of social research.* London: Longman.

Husserl, E. 1970. *The idea of phenomenology.* (Trans. W. Alstort & G. Nakhnikian.) The Hague: Martinus Nijhoff.

Hutcheon, L. 1986–87. The politics of postmodernism: parody and history. *Cultural Critique* **5**, 179–207.

Jackson, J. B. 1964. The meaning of landscape. *Kulturgeographi* **88**, 47–51.

Jackson, J. B. 1980. *The necessity of ruins.* Amherst, Mass.: University of Massachusetts Press.

Jackson, P. 1981. Phenomenology and social geography. *Area* **13**, 299–305.

Jackson, P. 1985. Urban ethnography. *Progress in Human Geography* **9**, 157–76.

Jackson, P. 1986. Social geography: the rediscovery of place. *Progress in Human Geography* **10**(1), 118–24.

Jackson, P. & S. J. Smith 1984. *Exploring social geography.* London: George Allen & Unwin.

Jacobs, J. 1961. *The death and life of great American cities.* New York: Vintage.

Jarrie, I. C. 1983. Realism and the supposed poverty of sociological theory. In *Epistemology, methodology and the social sciences,* R. S. Cohen & M. W. Wartofsky (eds), 107–21. Dordrecht & Boston: Reidel.

Jay, M. 1973. *The dialectical imagination.* Boston, and Toronto: Little, Brown & Company.

Jencks, C. 1981. *The language of post-modern architecture.* New York: Rizzoli.

Jencks, C. 1985. *Ornament and symbolism in post-modern architecture*. Paper presented to the Emily Carr College of Art and Design, Vancouver, BC.

Jencks, C. 1986. *What is post-modernism?* New York: St Martin's Press.

Jencks, C. & W. Chaitkin 1982. *Architecture today*. New York: Harry Abrams.

Jessop, B. 1982. *The capitalist state*. Oxford: Martin Robertson.

Johnson, H. B. and G. Pitzl 1981. Viewing and perceiving the rural scene: visualization in human geography. *Progress in Human Geography* **5**(2), 211–33.

Johnson, S. 1984. When home becomes your corporate headquarters. *Working Woman*, October, 75–7.

Johnston, R. J. 1983. *Geography and geographers: Anglo-American geography since 1945*, 2nd edn. London: Edward Arnold.

Johnston, R. J. 1984. The world is our oyster. *Transactions, Institute of British Geographers* **9**, 443–59.

Johnston, R. J. 1987. *Geography and geographers: Anglo-American human geography since 1945*, 3rd edn. London: Edward Arnold.

Kant, E. 1956. *Critique of practical reason*. (Trans. L. W. Beck.) Indianapolis: Bobbs Merrill.

Kant, E. 1964. *Critique of pure reason*. (Trans. N. Smith-Kemp.) New York: St Martin's Press.

Keat, R. & J. Urry 1975. *Social theory as science*. London: Routledge & Kegan Paul.

Keat, R. & J. Urry 1982. *Social theory as science*, 2nd edn. London: Routledge & Kegan Paul.

Keller, S. (ed.) 1981. *Building for women*. Lexington, Mass.: Lexington Books.

Kennedy, B. 1979. A naughty world. *Transactions of the Institute of British Geographers* NS **4**, 550–58.

Kiernan, M. 1982. Ideology and the precarious future of the Canadian planning profession. *Plan Canada* **22**(1), 14–24.

Kingston Community Project 1965a. *Kingston community project prospectus*. A project of the Students' Union for Peace Action (SUPA). Mimeograph. Kingston: Queen's University.

Kingston Community Project 1965b. *Kingston community project special report*. Meeting Poverty Series. Ottawa: Special Planning Secretariat, Department of State.

Klee, P. 1948. *On modern art*. London: Faber & Faber.

Klotz, H. (ed.) 1985. *Postmodern visions*. New York: Abbeville Press.

Knox, P. 1987. The social production of the built environment: architects, architecture and the post-modern city. *Progress in Human Geography* **11**, 354–78.

Kobayashi, A. 1983. *Emigration from Kaideima, Japan, 1885–1950: an analysis of community and landscape change*. Unpublished PhD dissertation, University of California at Los Angeles.

Kockelmans, J. J. 1967. *Phenomenology: the philosophy of Edmund Husserl and its interpretation.* Garden City, New York: Anchor Books, Doubleday & Company.

Kolakowski, L. 1968. *The alienation of reason.* New York: Doubleday.

Kosik, K. 1976. *Dialectics of the concrete.* Dordrecht and Boston: D. Reidel.

Kurzweil, E. 1980. *The age of structuralism.* New York: Columbia University Press.

Kuttner, R. 1985. The poverty of economics. *Atlantic Monthly* **255**(2), 74–84.

Kyburg, H. E. Jr. 1970. Conjunctivitis. In *Induction, acceptance, and rational belief*, M. Swain (ed.), 55–82. Dordrecht and Boston: D. Reidel.

LaCapra, D. 1983. *Rethinking intellectual history: texts, contexts, language.* London: Cornell University Press.

Lakatos, I. 1978. *The methodology of scientific research programmes.* Cambridge: Cambridge University Press.

Lang, R. & A. Armour 1980. *Environmental planning resourcebook.* Ottawa: Environment Canada.

Lash, H. 1976. *Planning in a human way.* Ottawa: Ministry of State for Urban Affairs.

Laxer, J. 1970. Americanisation of the Canadian student movement. In *Close the 49th Parallel etc*, I. Lumsden (ed.). Toronto: University of Toronto Press.

League for Social Reconstruction 1935. *Social planning for Canada.* Toronto: Thomas Nelson.

Le Corbusier 1933. *La Ville Radieuse.* Paris: L'Architecture d'Aujourd'hui.

Lefebvre, H. 1976. *The survival of capitalism.* London: Allison & Busby.

Leijonhufvud, A. 1973. Life among the Econ. *Western Economic Journal* **11**, 327–37.

Leontief, W. 1971. Theoretical assumptions and non–observed facts. *American Economic Review* **61**, 1–7.

Levesque, J.-M. 1985. Self–employment in Canada: a closer examination. *The Labour Force*, February, 91–105.

Lévi-Strauss, C. 1958. *Anthropologie Structurale.* Paris: Plon. (Trans. 1963.) *Structural anthropology.* New York: Basic Books.

Levitt, C. 1984. *Children of privilege. Student revolt in the sixties.* Toronto: University of Toronto Press.

Ley, D. 1974. *The black inner city as frontier outpost.* Washington, DC: Association of American Geographers, Monograph Series, No. 7.

Ley, D. 1978. Social geography and social action. In *Humanistic geography: problems and prospects*, D. Ley & M. Samuels (eds), 41–57. Chicago: Maaroufa/Croom Helm.

Ley, D. 1980. *Geography without man: a humanistic critique.* Research Paper 24, School of Geography, Oxford University.

Ley, D. 1981. Cultural/humanistic geography. *Progress in Human Geography* **5**(2), 249–57.

Ley, D. 1982. Rediscovering man's place. *Transactions of the Institute of British Geographers* NS **7**, 248–53.

Ley, D. 1983a. Cultural/humanistic geography. *Progress in Human Geography* **7**(2), 267–75.

Ley, D. 1983b. *A social geography of the city*. New York: Harper & Row.

Ley, D. 1984. *Gentrification in an urban system: examining alternative explanations*. Paper presented at the Annual Meeting of the American Association of Geographers, Washington, DC.

Ley, D. 1985a. Cultural/humanistic geography. *Progress in Human Geography* **9**(3), 415–23.

Ley, D. 1985b. Downtown or the suburbs? A comparative study of two Vancouver head offices. *The Canadian Geographer* **29**, 30–43.

Ley, D. 1986. *Modernism, post-modernism, and the struggle for place*. Paper presented to the seminar series, The Power of Place, Department of Geography, Syracuse University.

Ley, D. 1987. Styles of the times: liberal and neoconservative landscapes in inner Vancouver, 1968–1986. *Journal of Historical Geography* **13**, 40–56.

Ley, D. & G. Pratt 1983. Is philosophy necessary? *Geographical Analysis* **15**, 64–9.

Ley, D. & M. Samuels (eds) 1978. *Humanistic geography: prospects and problems*. Chicago and London: Maaroufa/Croom Helm.

Lindblom, C. E. 1959. The science of 'muddling through'. *Public Administration Review* **19**, 79–88.

Lindblom, C. E. 1965. *The intelligence of democracy*. New York: Free Press.

Lippi, M. 1979. *Value and materialism in Marx*. London: New Left Books.

Lowenthal, D. 1961. Geography, experience and imagination: towards a geographical epistemology. *Annals A.A.G.* **51**, 241–60.

Lowenthal, D. 1978. Finding valued landscapes. *Progress in Human Geography* **2**, 373–418.

Lowenthal, D. & M. Bowden 1975. *Geographies of the mind: essays in historical geosophy in honour of John Kirtland Wright*. New York: Oxford University Press.

Lowenthal, D. & H. Prince 1965. English landscape tastes. *The Geographical Review* **55**, 186–222.

Lynch, K. 1981. *A theory of good city form*. Cambridge: Massachusetts Institute of Technology.

McCarney, J. 1981. Review of A. Skillen's Ruling Illusions. *Radical Philosophy* **27**.

McCarthy, T. 1978. *The critical theory of Jürgen Habermas*. London: Macmillan.

McDowell, L. 1984. Towards an understanding of the gender division of urban space. *Environment and Planning D: Society and Space* **1**(1), 59–72.

McDowell, L. 1985. Some gloomy thoughts from Britain. *Women and Environments* **7**(2), 10–11.

McGee, T. G. 1984. *Circuits and networks of capital. The internationalization of the world economy and national urbanization*. Paper presented at the

Conference on Urban Growth and Economic Development in the Pacific Region, Taipei, Taiwan.

McGuiness, B. 1982. Freud and Wittgenstein. In *Wittgenstein and his times*, B. McGuiness (ed.), 427–43. Oxford: Basil Blackwell.

McHarg, I. L. 1969. *Plan with nature*. Philadelphia: Falcon Press for the National History Press.

Mackenzie, S. 1984. Catching up with ourselves: ideas on developing gender-sensitive theory in the environmental disciplines. *Women and Environments* **6**(3), 16–18.

Mackenzie, S. & D. Rose 1982. On the necessity for feminist scholarship in human geography. *The Professional Geographer* **34**(2), 220–23.

Mackenzie, S. & D. Rose 1983. Industrial change, the domestic economy and home life. In *Redundant spaces in cities and regions? Studies in industrial decline and social change*, J. Anderson, S. Duncan & R. Hudson (eds), 155–200. London: Academic Press.

McLoughlin, J. B. 1969. *Urban and regional planning: a systems approach*. London: Faber & Faber.

Mahon, R. 1984. *The politics of industrial restructuring – the Canadian textiles*. Toronto: University of Toronto Press.

Malecki, E. J. 1986. Technological imperatives and modern corporate strategy. In *Production, work, territory. The geographical anatomy of industrial capitalism*, A. J. Scott & M. Storper (eds), 67–79. Boston: Allen & Unwin.

Manion, J. 1983a. *Some alternative views on steel industry problems: a critique of steelmakers' perceptions and behaviour*. Waterloo: Department of Management Sciences, University of Waterloo.

Manion, J. 1983b. *Expansion and modernisation in the US steel industry, 1980–1982*. Waterloo: Department of Management Sciences, University of Waterloo.

Mannheim, K. 1946. Planning for freedom. In *Social change*, A. Etzioni & E. Etzioni (eds), 463–7. New York: Basic Books.

Martin, B. 1981. *A sociology of contemporary cultural change*. Oxford: Basil Blackwell.

Martines, L. 1980. *Power and imagination: city states in Renaissance Italy*. New York: Knopf.

Massey, D. 1978. Regionalism: some current issues. *Capital and Class* **6**, 106–25.

Massey, D. 1984. *Spatial divisions of labour: social structures and the geography of production*. London: Macmillan Education.

Massey, D. & J. Allen (eds) 1984. *Geography matters*. Cambridge: Cambridge University Press.

Massey, D. & R. A. Meegan 1978. Industrial restructuring versus the cities. *Urban Studies* **15**, 273–88.

Massey, D. & R. A. Meegan 1982. *The anatomy of job loss: the how, why and where of employment decline*. London: Methuen.

Mattera, P. 1985. *Off the books: the rise of the underground economy*. London: Pluto.

Mattick, P. Jr. 1986. *Social knowledge*. London: Hutchinson.

May, J. A. 1970. *Kant's concept of geography.* Toronto: University of Toronto Press.

Meek, R. L. 1976. *Studies in the labour theory of value,* 2nd edn. London: Lawrence & Wishart.

Meinig, D. 1976. The beholding eye: ten versions of the same scene. *Landscape Architecture* **66**, 47–54.

Meinig, D. (ed.) 1979. *The interpretation of ordinary landscapes: geographical essays.* New York: Oxford University Press.

Meinig, D. 1983. Geography as an art. *Transactions of the Institute of British Geographers* NS **8**, 314–28.

Meszaros, I. 1979. *The work of Sartre.* Vol. I: *Search for freedom.* Atlantic Highlands, New Jersey: Humanities Press.

Meszaros, I. 1987. Marx's 'social revolution' and the division of labour. *Radical Philosophy* **44**, 14–23.

Meyerson, M. 1956. Building the middle-range bridge for comprehensive planning. *Journal of th American Institute of Planners* **22**, 58–64.

Michelson, W. 1985. *From sun to sun: daily obligations and community structure in the lives of employed women and their families.* Totowa, New Jersey: Rowman & Allenheld.

Miles, A. 1985. Feminist radicalism in the 1980s. *Canadian Journal of Political and Social Theory* **9**(1–2), 16–39.

Millett, K. 1969. *Sexual politics.* New York: Avon.

Mills, C. W. 1971. *The sociological imagination.* New York: Oxford University Press.

Mirowski, P. 1984. The role of conservation principles in twentieth-century economic theory. *Philosophy of the Social Sciences* **14**, 46–73.

Mitchell, J. 1974. *Psychoanalysis and feminism.* New York: Pantheon.

Monk, J. & S. Hanson 1982. On not excluding half the human in human geography. *The Professional Geographer* **34**(1), 11–23.

Moos, A. & M. Dear 1986. Structuration theory in urban analysis: 1. Theoretical exegesis. *Environment and Planning A* **18**, 231–52.

Morgan, R. 1970. *Sisterhood is powerful: an anthology of writings from the women's liberation movement.* New York: Vintage.

Morris, A. E. J. 1979. *History of urban form.* London: George Godwin.

Morris, D. 1984. *The new city states.* Washington: Institute for Local Self-Reliance.

Mueller, H. 1982. The steel industry. In *The Internationalization of the American Economy,* M. J. Finger & T. D. Willet (eds). Annals of the American Academy of Political Science. Beverly Hills: Sage.

Mumford, L. 1967–70. *The myth of the machine: technics and human development.* New York: Harcourt, Brace & World.

Murgatroyd, L., M. Savage, D. Shapiro, J. Urry, S. Walby, A. Warde, with J. Mark-Lawson 1985. *Localities, class and gender.* London: Pion.

Napoleoni, C. 1978. Sraffa's 'tabula rasa'. *New Left Books* **112**, 75–7.

Natanson, M. 1963. *Philosophy of the social sciences.* New York: Random House.

Newman, J. & D. Crossfield 1968. *Grant application to the company of young Canadians*. Unpublished manuscript, Kingston.

Nicolaides, K. 1941. *The natural way to draw*. Boston: Houghton Mifflin.

Norton, W. 1987. Humans, land and landscape: a proposal for cultural geography. *Canadian Geographer* **31**, 21–30.

O'Connor, J. 1973. *The fiscal crisis of the state*. London: St James Press.

Ollman, B. 1976. *Alienation: Marx's conception of man in capitalist society*, 2nd edn. Cambridge: Cambridge University Press.

Olsson, G. 1979. Social science and human action or on hitting your head against the ceiling of language. In *Philosophy in geography*, S. Gale & G. Olsson (eds), 287–307. Dordrecht, Holland: Reidel.

Olsson, G. 1982. Epilogue: a ground for common search. In *A search for common ground*, P. Gould & G. Olsson (eds). London: Pion.

Olsson, G. 1983. Expressed impressions of impressed expressions. *Geographical Analysis* **15**, 60–64.

Oppong, C. 1983. Women's roles and conjugal family systems in Ghana. In *The changing position of women in family and society: a cross-national comparison*, E. Lupri (ed.), 331–43. Leiden: Brill.

Ornstein, M., A. M. Stevenson & A. P. Williams 1980. Region, class and political culture in Canada. *Canadian Journal of Political Science* **13**, 227–71.

Ozbekhan, H. 1968. Towards a general theory of planning. In *Perspectives of planning*, E. Jantsch (ed.), 47–158. Paris: OECD.

Pahl, R. 1984. *Divisions of labour*. Oxford: Basil Blackwell.

Palloix, C. 1975. The internationalization of capital and the circuit of social capital. In *International firms and modern imperialism*, H. Radice (ed.), 63–88. Harmondsworth: Penguin.

Palm, R. & A. Pred 1978. The status of women: a time-geographic view. In *An invitation to geography*, 2nd edn, D. Lanegran & R. Palm (eds), 99–109. New York: McGraw-Hill.

Papageorgiou, Y. 1982. Some thoughts about theory in social sciences. *Geographical Analysis* **14**, 340–46.

Paris, C. (ed.) 1982. *Critical readings in planning theory*. Oxford: Pergamon.

Parsons, T. 1949. *The structure of social action*, 2nd edn. New York: Free Press.

Paterson, J. 1983. *David Harvey's geography*. London: Croom Helm.

Peet, R. 1977. The development of radical geography in the United States. In *Radical geography*, R. Peet (ed.), 6–30. Chicago: Maaroufa.

Pickles, J. 1984. *Phenomenology, science and geography*. Cambridge: Cambridge University Press.

Pickles, J. 1986. Geographic theory and educating for democracy. *Antipode* **18**(2), 136–54.

Plunkett, T. J. 1968. *Urban Canada and its government. A study of municipal organisation*. Toronto: Macmillan.

Popper, K. R. 1944. *The open society and its enemies*, reprinted 1971. Princeton, NJ: Princeton University Press.

Poster, M. 1978. *Existential Marxism in postwar France.* Princeton: Princeton University Press.

Poster, M. 1984. *Foucault, Marxism and history.* Cambridge: Polity.

Postgate, D. & K. McRoberts 1980. *Quebec: social change and political crisis,* rev. edn. Toronto: McClelland & Stewart.

Powell, J. M. 1979. The haunting of Saloman's house. *The Australian Geographer* **14**, 327–41.

Pratt, G. 1984. *An appraisal of the incorporation thesis: housing tenure and political values in urban Canada.* Unpublished PhD dissertation, University of British Columbia.

Pred, A. R. 1983. Structuration and place: on the becoming of sense of place and structure of feeling. *Journal for the Theory of Social Behaviour* **13**, 45–68.

Pred, A. R. 1984a. Place as historically contingent process: structuration and the time-geography of becoming places. *Annals* A.A.G. **74**(2), 279–97.

Pred, A. R. 1984b. Structuration, biography formation and knowledge: observations on post growth during the late mercantile period. *Environment and Planning D.: Society and Space* **2**(3), 251–76.

Prince, H. 1961. The geographical imagination. *Landscape* **11**, 22–5.

Prince, H. 1971a. Questions of social relevance. *Area* **3**(3), 150–53.

Prince, H. 1971b. Real, imagined and abstract worlds of the past. *Progress in Geography* **3**, 1–86.

Quebec 1979. An Act respecting land use and development. *Bill 125.*

Quick, J. & P. O'Sullivan 1986. Military analysis of urban terrain. *The Professional Geographer* **38**(3), 286–90.

Rapoport, A. 1984. Culture and the urban order. In *The city in cultural context,* J. Agnew et al. (eds), 50–75. Boston: Allen & Unwin.

Rawls, J. 1971. *A theory of justice.* Cambridge, Mass.: Harvard University Press.

Reclus, E. 1899. *The ocean, atmosphere and life.* London: Chapman & Hall.

Redclift, N. & E. Mingione (eds) 1985. *Beyond employment: household, gender and subsistence.* Oxford: Basil Blackwell.

Reich, C. 1970. *The greening of America.* New York: Random House.

Relph, E. 1970. An inquiry into the relations between phenomenology and geography. *The Canadian Geographer* **14**(3), 193–201.

Relph, E. 1981. *Rational landscapes and humanist geography.* London: Croom Helm; Totowa: Barnes & Noble.

Relph, E. 1984. Seeing, thinking and describing landscapes. In *Environmental perception and behaviour,* T. F. Saarinen, D. Seamon & J. Sell (eds), 209–23. Research Paper 209. Chicago: Department of Geography, University of Chicago.

Roach, R. & B. Rosas 1972. Advocacy geography. *Antipode* **4**(2), 69–76.

Roberts, H. (ed.) 1981. *Doing feminist research.* Harmondsworth: Penguin.

Robbins, L. 1932. *An essay on the nature and significance of economic science,* 2nd edn. London: Macmillan.

Robinson, I. M. (ed.) 1972. *Decision making in urban planning.* Beverly Hills: Sage.

Robinson, J. V. 1962. *Economic philosophy.* Harmondsworth: Penguin.

Roncaglia, A. 1978. *Sraffa and the theory of prices.* London: John Wiley.

Rookmaaker, H. R. 1973. *Modern art and the death of a culture.* Downers Grove, Ill.: Inter-Varsity Press.

Rosenau, H. 1959. *The ideal city.* London: Routledge & Kegan Paul.

Roussopoulos, D. (ed.) 1982. *The city and radical social change.* Montreal: Black Rose Press.

Rowbotham, S. 1972. Women's liberation and the new politics. In *The body politic: women's liberation in Britain, 1969–1972*, M. Wandor (ed.), 3–20. London: Stage 1.

Rowbotham, S. 1973. *Woman's consciousness, man's world.* Harmondsworth: Penguin.

Rowbotham, S. 1983. *Dreams and dilemmas: collected writings.* London: Virago.

Rowbotham, S., L. Segal, & H. Wainwright 1979. *Beyond the fragments: feminism and the making of socialism.* London: Merlin.

Roweis, S. 1982. Urban planning in early and late capitalist societies. In *Urbanization and urban planning in capitalist society*, M. Dear & A. Scott (eds), 159–77. London and New York: Methuen.

Ruskin, J. 1856. *Modern painters*, 1904 edn. London: George Allen.

Ruskin, J. 1857. *The elements of drawing*, 1971 edn. New York: Dover.

Rutherford, P. (ed.) 1974. *Saving the Canadian city.* Toronto: University of Toronto Press.

Sack, R. D. 1980. *Conceptions of space in social thought.* London: Macmillan.

Sack, R. D. 1982. Realism and realistic geography. *Transactions of the Institute of British Geographers* NS **7**, 504–9.

Sack, R. D. 1983. Reply to R.A. Sayer. *Transactions of the Institute of British Geographers* NS **8**, 512–14.

Salmon, W. C. 1975. A third dogma of empiricism. In *Basic problems in methodology and linguistics. Part three of the proceedings of the fifth congress of logic, methodology, and philosophy of science*, R. E. Butts & J. Hintikka (eds), 149–66. Dordrecht and Boston: Reidel.

Salter, C. L. 1976. The role of landscape modification in revolutionary nation-building: the case of Mao's China. *The China Geographer* **4**, 41–59.

Salter, C. L. 1981. The new localism in China's cultural landscapes. *Landscape* **5**(3), 10–14.

Salter, C. L. & W. J. Lloyd 1977. *Landscape in literature.* Resource Paper for College Geography **76**(3). Washington: Association of American Geographers.

Samuels, M. S. 1978. Existentialism and human geography. In *Humanistic geography*, D. Ley & M. S. Samuels (eds), 22–40. Beckenham: Croom Helm.

Samuels, M. S. 1979. The biography of landscape: cause and culpability.

In *The interpretation of ordinary landscapes*, D. W. Meinig (ed.), 51–88. Oxford: Oxford University Press.

Santos, M. 1983. Geography in the late twentieth century: new roles for a threatened discipline. *Epistemology of Social Science*. UNESCO.

Sartre, J.-P. 1956. *Being and nothingness*. (Trans. H. Barnes.) New York: Philosophical Library.

Sartre, J.-P. 1963. *Search for a method*. (Trans. H. E. Barnes.) New York: Vintage.

Sartre, J.-P. 1971. Replies to structuralism: an interview with Jean-Paul Sartre. (Trans. R. d'Amico.) *Telos* 9, 110–16.

Sartre, J.-P. 1976. *Critique of dialectical reason*. (Trans. A. Sheridan-Smith.) London: New Left Books.

Saunders, P. 1981. *Social theory and the urban question*. London: Hutchinson.

Saunders, P. 1985. Review of *Method in social science: a realistic approach*. By A. Sayer 1984. London: Hutchinson. *Society and Space* 3(4), 489–95.

Savas, E. S. 1982. *Privatising the public sector: how to shrink government*. Chatham, NJ: Chatham House.

Sayer, A. 1978. Mathematical modelling in regional science and political economy: some comments. *Antipode* 10, 79–86.

Sayer, A. 1979a. Epistemology and conceptions of people and nature in geography. *Geoforum* 10(1), 19–44.

Sayer, A. 1979b. Theory and empirical research in urban and regional political economy: a sympathetic critique. *Working Paper 14: Urban and Regional Studies*. University of Sussex.

Sayer, A. 1982a. Explanation in economic geography: abstraction versus generalization. *Progress in Human Geography* 6(1), 68–88.

Sayer, A. 1982b. Misconceptions of space in social thought. *Transactions of the Institute of British Geographers* NS 7, 494–503.

Sayer, A. 1983. Reply to Robert Sack. *Transactions, Institute of British Geographers* 8, 508–11.

Sayer, A. 1984. *Method in social science: a realist approach*. London: Hutchinson.

Sayer, D. 1979. *Marx's method*. Hassocks: Harvester Press.

Schumacher, E. H. 1974. *Small is beautiful: a study of economics as if people mattered*. London: Arnold.

Schutz, A. 1962. *Collected papers I: the problems of social reality*, Vol. 1. (Ed. and introduced by Maurice Natason.) The Hague: Martinus Nijhoff.

Scott, A. J. 1984. Towards a theoretical human geography. *Environment and Planning D: Society and Space* 2(3), 119–21.

Sennett, R. & J. Cobb 1972. *The hidden injuries of class*. New York: Alfred A. Knopf.

Shaikh, A. 1978. An introduction to the history of crisis theories. In *US capitalism in crisis*, 219–41. New York: Union of Radical Political Economics.

Shmueli, E. 1973. Can phenomenology accommodate Marxism? *Telos* 17, 169–80.

Simmie, J. M. 1974. *Citizens in conflict: the sociology of town planning.* London: Hutchinson.

Simon, H. A. 1957. *Administrative behavior: a study of decision-making processes in administrative organizations.* New York: Free Press.

Slater, D. 1977. The poverty of modern geographical enquiry. In *Radical geography*, R. Peet (ed.), 40–57. Chicago: Maaroufa.

Smart, B. 1976. *Sociology, phenomenology and Marxian analysis.* London: Routledge & Kegan Paul.

Smith, D. 1971. Radical geography – the next revolution. *Area* 3(3), 152–7.

Smith, D. 1973. Alternative relevant professional roles. *Area* 5(1), 1–4.

Smith, D. C. 1976. The dual labour market theory: a Canadian perspective. *Research and Current Issues Series* 32. Kingston: Industrial Relations Centre, Queen's University.

Smith, N. 1979. Geography, science and post positivist modes of explanation. *Progress in Human Geography* 3(3), 356–83.

Smith, N. 1984. *Uneven development.* Oxford: Basil Blackwell.

Smith, S. J. 1981. Humanistic method in contemporary social geography. *Area* 13, 293–8.

Smith, S. J. 1984. Practicing humanistic geography. *Annals A.A.G.* 74, 353–74.

Snow, C. P. 1962. The moral un-neutrality of science. In *The New Scientist*, P. C. Obler & H. A. Estrin (eds). London: Anchor Books.

Spencer, M. 1977. History and sociology: an analysis of Weber's 'The city'. *Sociology* 11, 507–25.

Spragge, G. L. 1975. Canadian planners' goals: deep roots and fuzzy thinking. *Canadian Public Administration* 18(2), 216–34.

Squire, L. 1979. *Labour force, employment and labour markets in the course of economic development.* World Bank Staff Report 336. Washington, DC International Bank of Reconstruction and Development.

Sraffa, P. 1960. *Production of commodities by means of commodities.* Cambridge: Cambridge University Press.

Steedman, I. & P. W. Sweezy, 1981. *The value controversy.* London: Verso Editions and New Left Books.

Stelter, G. & A. F. J. Artibise 1977. *The Canadian city: essays in urban history.* Toronto: McClelland & Stewart.

Stephenson, D. 1974. The Toronto geographical expedition. *Antipode* 6(2), 98–101.

Stern, R. 1980. The doubles of post-modern. *Harvard Architecture Review* 1, 75–87.

Stimpson, C., E. Dixler, M. Nelson & K. Yatrakis (eds) 1981. *Women and the American city.* Chicago: University of Chicago Press.

Stoddart, D. R. 1986. *On geography and its history.* Oxford and New York: Basil Blackwell.

Storper, M. 1985. The spatial and temporal constitution of social action: a critical reading of Giddens. *Society and Space* 3(4), 407–24.

Straus, E. 1963. *The primary world of the senses.* London: Collier Macmillan.

Strunk, W. & E. B. White 1979. *The elements of style.* New York: Macmillan.

Susskind, L. 1983. *Resolving environmental regulatory disputes.* Cambridge, Mass.: Schenkman.

Szymanski, R. & J. Agnew, 1981. *Order and skepticism: human geography and the dialectic of science.* Resource publications in geography. Washington, DC: Association of American Geographers.

Taylor, E. G. R. 1930. *Tudor geography 1485–1583.* London: Methuen.

Thompson, E. P. 1968. *The making of the English working class.* Harmondsworth: Penguin.

Thompson, E. P. 1978. *The poverty of theory and other essays.* London: Merlin Press.

Thrift, N. 1983. On the determination of social action in space and time. *Environment and Planning D: Society and Space* **1**(1), 23–59.

Thrift, N. 1986. The geography of international economic disorder. In *A world in crisis? Geographical perspectives,* R. J. Johnston & P. J. Taylor (eds), 12–67. Oxford: Basil Blackwell.

Thwaites, A. T. & R. P. Oakey 1985. *The regional economic impact of technological change.* London: Frances Pinter.

Tilly, L. & J. Scott 1978. *Women, work and family.* New York: Holt, Rinehart & Winston.

Townson, M. 1984. Are computers destroying your home? *Goodwins* **2**(2), 16–19.

Tuan, Yi-Fu 1965. 'Environment' and 'world'. *Professional Geographer* **17**, 6–8.

Tuan, Yi-Fu 1971. Comments on geography and public policy. *The Professional Geographer* **15**, 181–92.

Tuan, Yi-Fu 1974. *Topophelia: a study of environmental perception, attitudes and values.* Englewood Cliffs, NJ: Prentice-Hall.

Tuan, Yi-Fu 1975. Place: an experiential perspective. *Geographical Review* **65**, 151–65.

Tuan, Yi-Fu 1977. *Space and place: the perspective of experience.* Minneapolis: University of Minnesota Press.

Tuan, Yi-Fu 1979. *Landscapes of fear.* Oxford: Basil Blackwell.

Tuan, Yi-Fu 1982. *Segmented worlds of self.* Minneapolis: University of Minnesota Press.

UNIDO 1984. Asian regional workshop on the integration of women in the industrial planning and development process. UNIDO, ID/WG. 424(4), 8 August.

van Ameringen, M. 1985. The restructuring of the Canadian auto industry. In *Canada and the new international division of labour,* D. Cameron & F. Houle (eds), 267–87. Ottawa: University of Ottawa Press.

Ventura, A. 1964. *Nobilita e Popolo Nella Societa Veneta del '400 e '500.* Bari.

Von Hayek, F. 1944. *The road to serfdom*. Chicago: University of Chicago Press.

Von Wright, G. 1958. Biographical sketch. In *Ludwig Wittgenstein, A Memoir*, N. Malcolm (ed.). London: Oxford University Press.

Wandor, M. (ed.) 1972. *The body politic: women's liberation in Britain 1969–1972*. London: Stage 1.

Ward, D. 1980. Environs and neighbours in the 'two nations': residential differentiation in mid-nineteenth century Leeds. *Journal of Historical Geography* **9** (1), 1–28.

Warnock, M. 1965. *The philosophy of Sartre*. London: Hutchinson.

Webber, M. & D. Rigby 1986. The rate of profit in Canadian manufacturing industry, 1950–1981. *Review of Radical Political Economics*.

Weisskopf, T. 1978. Marxist perspectives on cyclical crisis. In *US capitalism in crisis*, Union for Radical Political Economics (eds), 241–60. New York.

Wekerle, G. 1984. A woman's place is in the city. *Antipode* **16**(3), 11–19.

Wekerle, G., R. Peterson, & D. Morley (eds) 1980. *New space for women*. Boulder, Colorado: Westview Press.

Wellmer, A. 1972. *Critical theory of society*. Berlin: Herder & Herder.

Western, J. 1978. Knowing one's place: 'The coloured people' and the Group Areas Act in Cape Town. In *Humanistic geography*, D. Ley & M. Samuels (eds), 297–318. New York: Maaroufa/Methuen.

Western, J. 1986. The authorship of places: reflections on fieldwork in South Africa. *Syracuse Scholar* **7**, 4–17.

Westhues, K. 1975. Intergenerational conflict in the sixties. In *Prophecy and protest. Social movements in twentieth century Canada*, S. D. Clarke, J. P. Grayson & L. M. Grayson (eds), 387–408. Toronto: Gage.

Wharf, B. 1980. *Community work in Canada*. Toronto: McClelland & Stewart.

Wild, J. 1963. *Existence and the world of freedom*. Englewood Cliffs, NJ: Prentice-Hall.

Williams, M. 1983. The apple of my eye: Carl Sauer and historical geography. *Journal of Historical Geography* **9**(1), 1–28.

Williams, R. 1965. *The long revolution*. Harmondsworth: Penguin.

Williams, R. 1977. *Marxism and literature*. London: Oxford University Press.

Williams, R. 1980. *Problems in materialism and culture*. London: New Left Books.

Williams, R. 1982. *The sociology of culture*. New York: Schocken Books.

Williams, S. 1981. Realism, Marxism and human geography. *Antipode* **13**, 31–8.

Willis, P. 1979. *Learning to labour*. Aldershot: Gower.

Wilson, J. W. 1983. Planner in the middle: survival strategies for the store front planner. *Plan Canada* **22**(3,4), 75–85.

Winch, P. 1960. Nature and convention. In *The philosophy of society*, R. Beehler & A. R. Drengson (eds). London: Methuen.

Winch, P. 1979. Nature and convention. In *The philosophy of society*, R.

Beeher & A. R. Drengson (eds). London: Methuen.

Wisner, B. 1986. Geography: war or peace studies? *Antipode* **18**(2), 212–17.

Wittgenstein, L. 1958. *Philosophical investigations*. Oxford: Basil Blackwell.

Wolfe, J. M., I. Skelton & G. Drover 1980. Inner city real estate activity in Montreal: some institutional parameters. *The Canadian Geographer* **25**(4), 349–67.

Wolfe, T. 1975. *The painted word*. New York: Bantam Books.

Wolfe, T. 1981. *From Bauhaus to our house*. New York: Farrar, Straus, Giroux.

Women and Geography Study Group of the Institute of British Geographers 1984. *Geography and gender: an introduction to feminist geography*. London: Hutchinson.

Wootton, B. 1945. *Freedom under planning*. Chapel Hill: University of North Carolina Press.

Wright, E. O. 1978. *Class, crises and the state*. London: Verso.

Wright, E. O. 1979. The value of controversy and social research. *New Left Review* **116**, 53–82.

Wright, E. O. & L. Perrone 1977. Marxist class categories and income inequality. *American Sociological Review* **42**(1), 32–55.

Wright, F. L. 1934. Broadacre City model prepared and exhibited. In *Three quarters of a century of drawings*, A. Izzo, C. Gubitosi & F. L. Wright (eds). New York: Horizon Press.

Wright, J. K. 1947. Terrae incognitae: the place of imagination in geography. *Annals* A.A.G. **37**, 1–15.

Wright, J. K. 1966. *Human nature in geography*. Cambridge: Harvard University Press.

Yates, F. 1949. The Italian academies. Reprinted in *Renaissance and reform: the Italian contribution*. Volume 2, 1983, 6–30. London: Routledge & Kegan Paul.

Yates, F. 1964. *Giordano Bruno and the hermetic tradition*. London: Routledge & Kegan Paul.

Yates, F. 1972. *The Rosicrucian enlightenment*. London: Routledge & Kegan Paul.

Yates, F. 1979. *The occult philosophy in the Elizabethan age*. London: Routledge & Kegan Paul.

Young, R. M. 1979. Why are figures so significant? The role and critique of quantification. In *Demystifying social statistics*, J. Irvine, I. Miles & J. Evans (eds), 53–62. London: Pluto Press.

Zelinsky, W. 1970. Beyond the exponentials: the role of geography in the great transition. *Economic Geography* **46**, 499–535.

Zelinsky, W., J. Monk & S. Hanson 1982. Women and geography: a review and prospectus. *Progress in Human Geography* **6**(3), 317–66.

Index